ACTIVATED SLUDGE MODELS ASM1, ASM2, ASM2d AND ASM3

Edited by

IWA TASK GROUP ON MATHEMATICAL MODELLING FOR DESIGN AND OPERATION OF BIOLOGICAL WASTEWATER TREATMENT

Mogens Henze
Willi Gujer
Takashi Mino
Mark van Loosdrecht

Published by IWA Publishing in its Scientific and Technical Report series.

Published by IWA Publishing, Alliance House, 12 Caxton Street, London SW1H 0QS, UK

Telephone: +44 (0) 20 7654 5500; Fax: +44 (0) 20 7654 5555; Email: publications@iwap.co.uk
Web: **www.iwapublishing.com**

First published 2000
Reprinted 2002, 2007

Typeset in 10½/12 point New Caledonia by Perfect Page Publishing Services, London, UK
Cover design by Hilary Basing at Cato Street Studios, London, UK

British Library Cataloguing in Publication Data
A CIP catalogue record for this book is available from the British Library

ISBN: 978-1-900222-24-2
ISSN: 1025-0913

Preface

Modelling of activated sludge processes has become a common part of the design and operation of wastewater treatment plants. Today models are being used in design, control, teaching and research.

History

In 1982 the International Association on Water Pollution Research and Control (IAWPRC), as it was then called, established a Task Group on Mathematical Modelling for Design and Operation of Activated Sludge Processes. At that time modelling of activated sludge processes had been a discipline for about 15 years, most noticeably and reaching the most advanced level at the University of Cape Town, South Africa, by the research group headed by Professor G.v.R. Marais. The various models developed at that time had only little use, owing partly to lack of trust in the models, partly to the limitations in computer power and partly to the complicated way in which these models had to be presented in written form.

The first task

The aim for the Task Group was to create a common platform that could be used for future development of models for nitrogen-removal activated sludge processes. It was the aim to develop a model with a minimum of complexity. The result was the Activated Sludge Model No. 1, today known under many names: IAWPRC model, ASM1, IAWQ model, and so on.

The model outline was discussed at an IAWPRC Specialised Seminar at Kollekolle, Denmark, in 1985, and was published in 1987 in its final form in the IAWPRC Scientific and Technical Report Series as STR No. 1. The five years used for developing the model was spent in discussing with many researchers and practitioners in order to get a solid platform for the work and only to include details that could stand the test of time. What was presented was not only a model, but also a guideline for wastewater characterization and development of computer codes, plus a set of default values that since then has proved to give realistic model results with only minor changes in the parameters.

The ASM1 was well received and has been widely used as a basis for further model development. The direct use of the ASM1 for modelling has been almost nil, but ASM1 has been the core of numerous models with a number of supplementary details added in almost every case.

It was especially the matrix notation, which was introduced together with ASM1, that facilitated the communication of complex models and allowed the concentration of discussions on essential aspects of biokinetic modelling.

Biological phosphorus removal

At the time of publication of the ASM1, biological phosphorus removal was already being used in a (limited number) of full-scale treatment plants. The theoretical status of the processes was such that the Task Group at that time did not enter into the modelling of it. But from the mid-1980s to the mid-1990s the biological phosphorus removal processes grew very popular and at the same time the understanding of the basic phenomena of the process was increasing. Thus in 1995 the Activated Sludge Model no. 2 was published. This model included nitrogen removal and biological phosphorus removal. In 1994, when the ASM2 was finished, the role of denitrification in relation to biological phosphorus removal was still unclear, so it was decided not to include that element. However, the development in research was fast, and denitrifying PAOs (phosphorus-accumulating organisms) were needed for simulation of many results from research and practice. Because of this, the ASM2 model was expanded in 1999 into the ASM2d model, where denitrifying PAOs were included.

Although the models might not have been heavily needed for nitrogen removal processes, the complexity of the combined nitrogen and phosphorus removal processes makes the models important for design and control purposes.

New platform

The models have grown more complex over the years, from ASM1, including nitrogen removal processes, to ASM2, including biological phosphorus removal processes and to ASM2d including denitrifying PAOs. In 1998 the Task Group decided to develop a new modelling platform, the ASM3, in order to create a tool for use in the next generation of activated sludge models. The ASM3 is based on recent developments in the understanding of the activated sludge processes, among which are the possibilities of following internal storage compounds, which have an important role in the metabolism of the organisms.

Benefit from the models

The major impact of the ASM model family has been based upon three facts. The first is the common language that modellers speak when using the concepts, the nomenclature and the matrix notation of the ASMs. This has created a strong model development over the past 15 years, which would probably not have been the case if all the modellers had used their own concepts, notation and platforms.

The second is the organizing effect of working with a model. This has helped researchers to achieve more efficient experimental designs and helped treatment plant operators to better understand and organize the information available at their plants – and in many cases to spot errors in available information. The third is that the models have served as guidance for research. By demonstrating where research was needed, focus has been put on certain details, for example wastewater characterization, out of which much interesting research has grown.

Simulation programs

The ASM1 and ASM2 models, or ASM-based models, are included in most of today's commercial and non-commercial simulation programs. Thus it is easy to get access to, and use the models for various purposes.

Future

This report has been produced to give a total overview of the ASM model family status at the start of 2000 and to give to the reader easy access to the different models in their original versions. It is the hope of the present Task Group that this may facilitate the use of the models and their future development.

During the years the members of the Task Group have changed. This reflects the development in research over the years and the wish to develop the models further. The ASM3 is not the final or 'general model' for activated sludge. Like ASM1, it is a structure and a platform for further development. Many modellers are looking for the 'ultimate general model' for activated sludge systems. Experience over the past 15 years shows that new development comes fast and the 'general models' have a short half-life. Thus for the future development of ASMs, suggestions, experience and discussion points will be well received if the readers and users wish to share their ups and downs in modelling with members of the Task Group.

Mogens Henze
Willi Gujer
Mark van Loosdrecht
Takashi Mino

Contents

List of Task Group members

IAWPR/IAWPRC Task Group 1982–1990

Mogens Henze (Chairman), *Technical University of Denmark, Denmark*
C.P. Leslie Grady, *Clemson University, USA*
Willi Gujer, *Swiss Federal Institute for Aquatic Science and Technology, Zürich, Switzerland*
Gerrit v.R. Marais, *University of Cape Town, South Africa*
Tomonori Matsuo, *University of Tokyo, Japan*

IAWPRC/IAWQ Task Group 1990–1996

Mogens Henze (Chairman), *Technical University of Denmark, Denmark*
Willi Gujer, *Swiss Federal Institute for Aquatic Science and Technology, Zürich, Switzerland*
Takashi Mino, *University of Tokyo, Japan*
Tomonori Matsuo, *University of Tokyo, Japan*
Mark C. Wentzel, *University of Cape Town, South Africa*
Gerrit v.R. Marais, *University of Cape Town, South Africa*

IAWQ/IWA Task Group since 1996

Mogens Henze (Chairman), *Department of Environmental Science and Engineering, Technical University of Denmark, 2800 Lyngby, Denmark (mh@imt.dtu.dk)*
Willi Gujer, *Swiss Federal Institute for Aquatic Science and Technology, 8600 Dübendorf, Switzerland (gujer@eawag.ch)*
Mark van Loosdrecht, *Delft University of Technology, Julianalaan 67, 2628 BC Delft, The Netherlands (mark.vl@stm.tudelft.nl)*
Takashi Mino, *University of Tokyo, 7-3-1, Hongo, Bunkyo-ku, Tokyo, Japan (mino@env1st.t.u-tokyo.ac.jp)*

ACTIVATED SLUDGE MODEL NO. 1

by

IAWPRC TASK GROUP ON MATHEMATICAL MODELLING FOR DESIGN AND OPERATION OF BIOLOGICAL WASTEWATER TREATMENT

M. Henze, *Denmark*
C.P.L. Grady, Jr, *USA*
W. Gujer, *Switzerland*
G. v. R. Marais, *South Africa*
T. Matsuo, *Japan*

Contents

Introduction

MODELLING is an inherent part of the design of a wastewater treatment system, regardless of the approach used. At the fundamental level, a design model may be merely conceptual; that is the engineer reduces the complex system with which he is dealing to a conceptual image of how it functions. That image then determines the design approach employed. Often, however, the engineer recognizes that the conceptual model alone does not provide sufficient information for design and thus he constructs a physical model, such as a lab-scale reactor or a pilot plant, upon which various design ideas can be tested. Given sufficient time for testing, such an approach is entirely satisfactory. However, the engineer may find that limitations of time and money prevent exploration of all potentially feasible solutions. Consequently, he often turns to the use of mathematical models to define further the feasible design space. He may devise empirical models which incorporate a statistical approach to mimic the end results obtained by studies on the physical model, or if his conceptual understanding expands sufficiently, he may attempt to formulate mechanistically based models which seek to account for the major events occurring within the system itself. These mechanistic models are the more powerful because they allow extrapolation of the design space to conditions beyond that experienced on the physical model. In this way, many potentially feasible solutions may be evaluated quickly and inexpensively, thereby allowing only the more promising ones to be selected for actual testing in the physical model.

The modelling of biological wastewater treatment systems has passed through the above sequence of events several times: first, for the removal of organic matter only; second, for nitrification; and third, for nitrogen removal by biological denitrification. Wastewater treatment practice has now progressed to the point where all of these can be accomplished in a single sludge system. Because of the interactions within such systems, the mathematical models depicting them are quite complex, which has detracted from their use. This is unfortunate because it is with such complex systems that the engineer has the most to gain from the use of mathematical models.

Realizing the benefits to be derived from mathematical modelling, while recognizing the reluctance of many engineers to use it, the International Association on Water Pollution Research and Control (IAWPRC) formed a task group in 1983 to promote the development of, and facilitate the application of, practical models to the design and operation of biological wastewater treatment systems. The goal was first to review existing models and second to reach a consensus concerning the simplest one having the capability of realistic predictions of the performance of single sludge systems carrying out carbon oxidation, nitrification, and denitrification. The model was to be presented in a way that made clear the processes incorporated into it and the procedures for its use. This report is the result of the group's efforts. Many research papers have influenced the thinking of the task group, but space limitation makes it impossible to deal even cursorily with each contribution. Accordingly, the group chose to cite only those papers that consolidate work of previous investigators or those individual contributions which present important new concepts. Only a basic outline can be presented without detailed supporting evidence for the choices made; the reader should refer to the cited papers for detailed information. The deliberations of the group also led its members to propose some modifications to existing models which they believe should improve predictive quality.

Method of model presentation

SIMULATION of activated sludge system behaviour, incorporating phenomena such as carbon oxidation, nitrification and denitrification, must necessarily account for a large number of reactions between a large number of components. To be mathematically tractable while providing realistic predictions, the reactions must be representative of the most important fundamental processes occurring within the system. In this context the term *process* is used to mean a distinct event acting upon one or more system components. Furthermore, the model should quantify both the kinetics (rate–concentration dependence) and the stoichiometry (relationship that one component has to another in a reaction) of each process. Identification of the major processes and selection of the appropriate kinetic and stoichiometric expressions for each are the major conceptual tasks during development of a mathematical model. Consequently, most of this report will concern them.

Format and notation

One problem often associated with papers presenting models describing complex systems is that it is difficult to follow the development of the author's ideas. In particular, it is often difficult to trace all the interactions of the system components. The task group concluded that a matrix format, based on the work of Peterson (1965), for presentation of the model offered the best opportunity for overcoming this problem while conveying the maximum amount of information. Furthermore, they felt that the notation recommended by a previous task group (Grau *et al*., 1982) should be used. An illustration will introduce the matrix format and the notation.

Consider the situation in which heterotrophic bacteria are growing in an aerobic environment by utilizing a soluble substrate for carbon and energy. In one simple conceptualization of this situation, two fundamental processes occur: the biomass increases by cell growth and decreases by decay. Other events, such as oxygen utilization and substrate removal, also occur, but these are not considered to be fundamental because they result from biomass growth and decay and are coupled to them through the system stoichiometry. The simplest model of this situation must consider the concentrations of three components: biomass, substrate, and dissolved oxygen. The matrix incorporating the fate of these three components in the two fundamental processes is shown in Table 1.

The first step in setting up the matrix is to identify the components of relevance in the model. In this scenario these are biomass, substrate and dissolved oxygen, which are listed across the top of Table 1 by symbol and across the bottom by name and units. In conformity with IAWPRC nomenclature (Grau *et al*., 1982), insoluble constituents are given the symbol X and the soluble components S. Subscripts are used to specify individual components: B for biomass, S for substrate and O for oxygen. The index i is assigned to each component. In this case, i ranges from 1 to 3 for the three compounds in this simple model.

The second step in developing the matrix is to identify the biological processes occurring in the system; i.e. the conversions or transformations which affect the components listed. Only two processes are included in this example: aerobic growth of biomass and its loss by decay. These processes are listed in the leftmost column of the matrix. The index j is assigned to each process; in this case, $j = 1$ or 2.

The kinetic expressions or rate equations for each process are recorded in the rightmost column of the matrix in the appropriate row. Process rates are denoted by ρ_j, where j corresponds to the process as numbered in the leftmost column. If we were to use the simple Monod–Herbert (Herbert, 1958) model for this situation the rate expressions would be those in Table 1. The Monod equation, ρ_1, says that growth of biomass is proportional to biomass concentration in a first order manner and to substrate concentration in a mixed order manner. The Herbert expression, ρ_2, states that biomass decay is first order with respect to biomass concentration. The kinetic parameters used in the rate expressions are defined in the lower right corner of the table.

The elements within the matrix comprise the stoichiometric coefficients, ν_{ij}, which set out the mass relationships between the components in the individual processes. For example, growth of biomass (+1) occurs at the expense of soluble substrate ($-1/Y$); oxygen is utilized in the metabolic process [$-(1-Y)/Y$]. The coefficients, ν_{ij}, are greatly simplified by working in consistent units. In this case, all organic constituents have been expressed as equivalent amounts of chemical oxygen demand (COD); likewise, oxygen is expressed as negative oxygen demand. The

Table 1. **Process kinetics and stoichiometry for heterotrophic bacterial growth in an aerobic environment**

Continuity →

Mass Balance ↓

Component → i / j Process ↓	1 X_B	2 S_S	3 S_O	Process Rate, ρ_j $[ML^{-3}T^{-1}]$
1 Growth	1	$-\frac{1}{Y}$	$-\frac{1-Y}{Y}$	$\frac{\hat{\mu}S_S}{K_S+S_S}X_B$
2 Decay	−1		−1	bX_B
Observed Conversion Rates $ML^{-3}T^{-1}$	$r_i=\sum_j r_{ij}=\sum_j \nu_{ij}\rho_j$			Kinetic Parameters:
Stoichiometric Parameters: True growth yield: Y	Biomass [M(COD) L^{-3}]	Substrate [M(COD) L^{-3}]	Oxygen (negative COD) [M(−COD) L^{-3}]	Maximum specific growth rate: $\hat{\mu}$; Half-velocity constant: K_S; Specific decay rate: b

sign convention used in the matrix is negative for consumption and positive for production. All stoichiometric coefficients are defined in the lower left corner of the table.

Use in mass balances

Within a system, the concentration of a single component may be affected by a number of different processes. An important benefit of the matrix representation is that it allows rapid and easy recognition of the fate of each component, which aids in the preparation of mass balance equations. This may be seen by moving down the column representing a component, which is why the arrow marked 'Mass Balance' is placed at the lefthand side. The basic equation for a mass balance within any defined system boundary is:

$$\text{Input}-\text{Output}+\text{Reaction}=\text{Accumulation} \tag{1}$$

The input and output terms are transport terms and depend upon the physical characteristics of the system being modelled. The system reaction term, r_i, is obtained by summing the products of the stoichiometric coefficients ν_{ij} and the process rate expression ρ_j for the component i being considered in the mass balance:

$$r_i=\sum_j \nu_{ij}\rho_j \tag{2}$$

For example the rate of reaction, r, for biomass, X_B, at a point in the system would be:

$$r_{X_B}=\frac{\hat{\mu}S_S}{K_S+S_S}X_B-bX_B \tag{3}$$

for soluble substrate, S_S it would be:

$$r_{S_S}=-\frac{1}{Y}\frac{\hat{\mu}S_S}{K_S+S_S}X_B \tag{4}$$

for dissolved oxygen, S_O it would be:

$$r_{S_O}=-\left(\frac{1-Y}{Y}\right)\frac{\hat{\mu}S_S}{K_S+S_S}X_B-bX_B \tag{5}$$

To create the mass balance for each component within a given system boundary (e.g. a completely mixed reactor), the conversion rate would be combined with the appropriate advective (flow) terms for the particular system. These terms have not been shown here because the purpose of the example was to demonstrate how the matrix is used to define the fundamental reactions regardless of the system configuration. It should be emphasized, however, that the modelling of a particular physical system requires definition of the system boundary with the associated advective terms.

Continuity check

Another benefit of the matrix is that continuity may be checked by moving across the matrix, provided consistent units have been used because then the sum of the stoichiometric coefficients must be zero. This can be demonstrated by considering the decay process. Recalling that oxygen is negative COD so that its coefficient must be multiplied by −1, all COD lost from the biomass because of decay must be balanced by oxygen utilization. Similarly, for the growth process, the substrate COD lost from solution due to growth minus the amount converted into new cells must equal the oxygen used for cell synthesis.

Model incorporating carbon oxidation, nitrification and denitrification

BECAUSE of the long solids retention times (SRTs) and low specific growth rates incorporated into the design of most biological wastewater treatment systems, differences in effluent soluble biodegradable substrate concentration between different system configurations are generally small. Conversely, large differences in activated sludge concentrations and electron acceptor (either oxygen or nitrate) requirements are common. Furthermore, good design practice requires that a sufficient quantity of electron acceptor be supplied in response to both real time and space–time (location) dependent changes in demand, and that final settlers and sludge return systems be capable of handling all anticipated concentrations of solids. This suggests that models depicting substrate removal are important more for their impact upon activated sludge concentrations and electron acceptor requirements than for their ability to predict effluent substrate concentration. Consequently, primary consideration was given by the task group to prediction of activated sludge concentrations during selection of process stoichiometry and to estimation of electron acceptor requirements during development of the process rate expressions.

The development of a mathematical model involves compromises to balance conflicting needs. On the one hand, a model must incorporate the major events occurring within a system in a manner which is consistent with established knowledge about that system. On the other hand, the model equations must be solvable with a reasonable degree of effort. Difficulty of solution increases markedly as the number of processes increases. In addition, the more closely the process rate expressions reflect reality, the more complicated they are likely to be. A modeller should include only those processes which are essential to a realistic solution and must select rate expressions for them that allow the use of simplified solution techniques without detracting from the applicability of the results. In many cases, this may require the selection of greatly simplified rate expressions. Although such rate expressions may not depict perfectly the actual events occurring within a system, they can be used satisfactorily as long as they mimic well the outcome of those events. In selecting the processes and rate expressions to be included in the model presented here, the task group focused on the major events and selected the simplest rate expressions consistent with them.

Within this context, it should be noted that the task group employed the concept of switching functions to turn process rate equations on and off as environmental conditions are changed. This was particularly necessary for processes that depend upon the type of electron acceptor present. For example the bacteria which are responsible for nitrification are capable of growth only under aerobic conditions and their rate of growth will fall to zero as the dissolved oxygen concentration approaches zero, regardless of the concentration of their energy yielding substrate. This can be modelled by including a dissolved oxygen 'switch' in the process rate equations. The oxygen switching function adopted by the task group was:

$$\frac{S_O}{K_O+S_O} \qquad (6)$$

where S_O is the concentration of dissolved oxygen.

The selection of a small value for K_O means that the value of the switching function is near unity for moderate dissolved oxygen (DO) concentrations but decreases to zero as the DO concentration approaches zero. The fact that the function is mathematically continuous helps to eliminate problems of numerical instability which can occur during simulations with models which include rate equations that are switched on and off discontinuously. Similarly, processes which occur only when dissolved oxygen is absent may be turned on by a switching function of the form:

$$\frac{K_O}{K_O+S_O} \qquad (7)$$

It will be recalled that predictions of activated sludge concentration, rather than the concentrations of soluble constituents in the vessels, and electron acceptor requirements were the primary focus of the task group in the development of the model. Nevertheless, it is apparent that the values of switching constants like K_O will influence those predictions even though functions (6) and (7) were chosen more for their mathematical convenience than conformity to any fundamental rate laws. Consequently, care should be taken in the selection of the values for switching constants to ensure that model predictions are not biased.

Conceptual model

A matter that has been the cause of confusion and to a certain extent has inhibited the development of activated sludge theory is the lack of a consistent measure of the concentration of organic material in wastewater. Three measures have gained acceptance and are widely used: biochemical oxygen demand (BOD), total organic carbon (TOC), and chemical oxygen demand (COD). Of these we believe that COD is undoubtedly the superior measure because it alone provides a link between electron equivalents in the organic substrate, the biomass and the oxygen utilized (Gaudy and Gaudy, 1971). Furthermore, mass balances can be made in terms of COD. Consequently, the concentrations of all organic materials, including biomass, are in COD units in the following model.

The organic matter in a wastewater may be subdivided into a number of categories (McKinney and Ooten, 1969; Dold *et al.*, 1980). The first important subdivision is based on biodegradability.

Non-biodegradable organic matter is biologically inert and passes through an activated sludge system unchanged in form. Two fractions, depending on their physical state, can be identified: soluble and particulate. Inert soluble organic matter, S_I, leaves the system at the same concentration that it enters. Inert suspended organic matter, X_I, becomes enmeshed in the activated sludge and is removed from the system through sludge wastage. Because the waste sludge flow rate is smaller than the system inflow rate, a mass balance requires the concentration of X_I in the system to be higher than in the influent.

Biodegradable organic matter may be divided into two fractions: readily biodegradable and slowly biodegradable. For purposes of modelling, the readily biodegradable material, S_S, is treated as if it were soluble, whereas the slowly biodegradable material, X_S, is treated as if it were particulate. It should be recognized, however, that some slowly biodegradable material may actually be soluble. The readily biodegradable material consists of relatively simple molecules that may be taken in directly by heterotrophic bacteria and used for growth of new biomass. A portion of the energy (COD) associated with the molecules is incorporated into the biomass, whereas the balance is expended to provide the energy needed for the synthesis. The electrons associated with that portion are transferred to the exogenous electron acceptors (oxygen or nitrate). In contrast, the slowly biodegradable material, consisting of relatively complex molecules, must be acted upon extracellularly and converted into readily biodegradable substrate before it can be used. It is assumed that conversion of slowly biodegradable substrate into the readily biodegradable form (hydrolysis) involves no energy utilization and thus there is no utilization of electron acceptor associated with it.

The specific rate of hydrolysis of slowly biodegradable substrate is usually considerably lower than the specific rate of utilization of readily biodegradable substrate, so that it becomes the rate-limiting factor in the growth of biomass when X_S alone is present as substrate. Furthermore, the rate of hydrolysis is lower under anoxic conditions (only nitrate available as the terminal electron acceptor) than under aerobic conditions and is apparently completely stopped under anaerobic conditions (neither nitrate nor oxygen are present) (Van Haandel *et al.*, 1981). The division of substrate into two forms provides a built-in lag in uptake of electron acceptor which allows space–time dependent variations in oxygen and nitrate utilization to be modelled.

Heterotrophic biomass is generated by growth on readily biodegradable substrate under either aerobic or anoxic conditions, but is assumed to stop under anaerobic conditions. Biomass is lost by decay, which incorporates a large number of mechanisms including endogenous metabolism, death, predation and lysis. For reasons to be explained later, decay is assumed to result in the conversion of biomass into slowly biodegradable substrate and particulate products, X_P, which are inert to further biological attack (Dold *et al.*, 1980). The latter are similar in concept to the endogenous mass of McKinney and Ooten (1969) and act to reduce the viability of the suspended solids in a bioreactor. The loss of biomass by decay is assumed to occur at a rate which is independent of the nature or concentration of the electron acceptor present, but the conversion of the resultant slowly biodegradable substrate to a form that can be used for regrowth of new cells is influenced by the nature of the electron acceptor as discussed in the preceding paragraph.

Nitrogenous matter in a wastewater, like carbonaceous matter, can be divided into two categories: non-biodegradable and biodegradable, each with further subdivisions. With respect to the non-biodegradable fraction, the particulate portion is that associated with the non-biodegradable particulate COD; the soluble portion is usually negligibly small and is not incorporated into the model. The biodegradable nitrogenous matter may be subdivided into: ‘ammonia’ (both the free compound and its salts), S_{NH}; soluble organic nitrogen, S_{ND}; and particulate organic nitrogen, X_{ND}. Particulate organic nitrogen is hydrolysed to soluble organic nitrogen in parallel with hydrolysis of slowly biodegradable organic matter. The soluble organic nitrogen is acted on by heterotrophic bacteria and converted to ammonia nitrogen. The ammonia nitrogen serves as the nitrogen supply for synthesis of heterotrophic biomass and as the energy supply for growth of autotrophic nitrifying bacteria. For simplicity, the autotrophic conversion of ammonia nitrogen to nitrate nitrogen is considered to be a single step process which requires oxygen. The nitrate formed may serve as terminal electron acceptor for heterotrophic bacteria under anoxic conditions, yielding nitrogen gas. Cell decay of either autotrophic or heterotrophic biomass leads to release of particulate organic nitrogen which can re-enter the cycle.

Both heterotrophic and autotrophic biomass may be present in the wastewater itself, thereby having a strong effect upon system performance. However, the prevalence and intensity of this occurrence is still unknown and thus it was not considered by the task group in developing the model. It should be noted, however, that the only change required for its inclusion would be the addition of input terms to the appropriate mass balance equations.

Components in mathematical models

The components in the model are shown across the top and bottom of Table 2. Soluble inert and particulate inert organic matter, S_I and X_I, are not involved in any conversion processes and thus their columns (i = 1 and 3, respectively) contain no stoichiometric coefficients. Nevertheless, they are included because they are important to the performance of the process. Soluble inert organic matter contributes to the effluent COD. Particulate inert organic matter becomes a part of the volatile suspended solids in the activated sludge system. As discussed earlier, all organic constituents, including particulate ones, are expressed in COD units and that is reflected in Table 2.

Moving down the i = 2 column, it can be seen that readily biodegradable substrate, S_S, is removed by growth of heterotrophic bacteria under either aerobic or anoxic conditions and is formed by hydrolysis of particulate organic matter entrapped in the biofloc. The i = 4 column reveals that slowly biodegradable substrate X_S, is removed by hydrolysis but is formed by decay of both heterotrophic and autotrophic biomass. In other words, decay results in the transformation of cell material into slowly biodegradable substrate. This will be discussed further later.

The columns where i = 5 and 6 represent the biomass in the system, with $X_{B,H}$ denoting the heterotrophic biomass and $X_{B,A}$ the autotrophic biomass. Moving down the i = 5 column reveals that heterotrophic biomass can be formed by growth under either aerobic or anoxic conditions. It is destroyed by decay. As seen in the i = 6 column, growth of the autotrophs only occurs under aerobic conditions. They, too, are destroyed by decay.

The i = 7 column contains the particulate products arising from biomass decay, X_P (Kountz and Forney, 1959; McKinney and Ooten, 1969). As far as the process kinetics and stoichiometry are concerned, it is formed by decay of both heterotrophic and autotrophic biomass, but is not destroyed. In actuality, this fraction of biomass is probably not completely inert to biological attack (Obayashi and Gaudy, 1973). However, its rate of destruction is so low that for all practical purposes it appears inert within the SRTs normally encountered in activated sludge systems. Incorporation of this component in the model is one way of accounting for the fact that not all biomass in an activated sludge system is active (Weddle and Jenkins, 1971).

The volatile solids concentration (in COD units) in the activated sludge system is the sum of the five particulate terms: X_S, $X_{B,H}$, $X_{B,A}$, X_P, and X_I. An appropriate conversion factor can be applied to convert from COD units to volatile suspended solids units.

The i = 8 column contains the concentration of DO, S_O, in the reactor. The processes included in the matrix only act to remove oxygen from solution and none are given for its addition; i.e. the matrix includes only biological processes. In order to simulate variations in DO concentration, appropriate process rate expressions for oxygen transfer would have to be included with the transport terms when writing the mass balance equation for oxygen. Even if those terms are not included, the information in the i = 8 column can still be used to calculate the quantity of oxygen which must be supplied to meet the metabolic needs of the bacteria. Moving down this column reveals that oxygen utilization is associated only with aerobic growth of the heterotrophic and autotrophic biomass. None is associated with microbial decay. This differs from the more traditional approach (Grady and Lim, 1980). Decay is assumed to result in the release of slowly biodegradable substrate which is recycled back to soluble substrate and used for more cell growth. Thus the oxygen utilization normally associated directly with decay is calculated as if it occurs indirectly from growth of new biomass on released substrate (Dold *et al.*, 1980). The net loss of biomass associated with decay results from the fact that the heterotrophic yield is less than unity, so that the amount of new biomass grown from released substrate must always be less than the amount of biomass lost. The 4.57 term in the stoichiometric coefficient for aerobic growth of autotrophs is the theoretical oxygen demand associated with the oxidation of ammonia nitrogen to nitrate nitrogen.

The other electron acceptor included in the model is nitrate nitrogen, S_{NO}, which is produced by aerobic growth of the autotrophic bacteria and removed during anoxic growth of the heterotrophic biomass, as can be seen by moving down the i = 9 column. Although nitrite nitrogen is an intermediate formed during nitrification, for simplicity in modelling it has been assumed that nitrate is the only oxidized form of nitrogen present. The factor 2.86 in the stoichiometric coefficient for anoxic growth of the heterotrophic biomass is the oxygen equivalence for conversion of nitrate nitrogen to nitrogen gas (N_2) and is included

Table 2 Process kinetics and stoichiometry for carbon oxidation, nitrification, and dentrification

j	Component → i / Process ↓	1 S_I	2 S_S	3 X_I	4 X_S	5 $X_{B,H}$	6 $X_{B,A}$	7 X_P	8 S_O	9 S_{NO}	10 S_{NH}	11 S_{ND}	12 X_{ND}	13 S_{ALK}	Process Rate, ρ_j $[ML^{-3}T^{-1}]$
1	Aerobic growth of heterotrophs		$-\frac{1}{Y_H}$			1			$-\frac{1-Y_H}{Y_H}$		$-i_{XB}$			$-\frac{i_{XB}}{14}$	$\hat{\mu}_H\left(\frac{S_S}{K_S+S_S}\right)\left(\frac{S_O}{K_{O,H}+S_O}\right)X_{B,H}$
2	Anoxic growth of heterotrophs		$-\frac{1}{Y_H}$			1				$-\frac{1-Y_H}{2.86Y_H}$	$-i_{XB}$			$\frac{1-Y_H}{14\cdot 2.86Y_H}-i_{XB}/14$	$\hat{\mu}_H\left(\frac{S_S}{K_S+S_S}\right)\left(\frac{K_{O,H}}{K_{O,H}+S_O}\right)\times\left(\frac{S_{NO}}{K_{NO}+S_{NO}}\right)\eta_g X_{B,H}$
3	Aerobic growth of autotrophs						1		$-\frac{4.57-Y_A}{Y_A}$	$\frac{1}{Y_A}$	$-i_{XB}-\frac{1}{Y_A}$			$-\frac{i_{XB}}{14}-\frac{1}{7Y_A}$	$\hat{\mu}_A\left(\frac{S_{NH}}{K_{NH}+S_{NH}}\right)\left(\frac{S_O}{K_{O,A}+S_O}\right)X_{B,A}$
4	'Decay' of heterotrophs				$1-f_P$	-1		f_P					$i_{XB}-f_P i_{XP}$		$b_H X_{B,H}$
5	'Decay' of autotrophs				$1-f_P$		-1	f_P					$i_{XB}-f_P i_{XP}$		$b_A X_{B,A}$
6	Ammonification of soluble organic nitrogen										1	-1		$\frac{1}{14}$	$k_a S_{ND} X_{B,H}$
7	'Hydrolysis' of entrapped organics		1		-1										$k_h\frac{X_S/X_{B,H}}{K_X+(X_S/X_{B,H})}\left[\left(\frac{S_O}{K_{O,H}+S_O}\right)+\eta_h\left(\frac{K_{O,H}}{K_{O,H}+S_O}\right)\left(\frac{S_{NO}}{K_{NO}+S_{NO}}\right)\right]X_{B,H}$
8	'Hydrolysis' of entrapped organic nitrogen											1	-1		$\rho_7(X_{ND}/X_S)$
	Observed Conversion Rates $[ML^{-3}T^{-1}]$	$r_i=\sum_j \nu_{ij}\rho_j$													
	Stoichiometric Parameters: Heterotrophic yield: Y_H; Autotrophic yield: Y_A; Fraction of biomass yielding particulate products: f_P; Mass N/Mass COD in biomass: i_{XB}; Mass N/Mass COD in products from biomass: i_{XP}	Soluble inert organic matter $[M(COD)L^{-3}]$	Readily biodegradable substrate $[M(COD)L^{-3}]$	Particulate inert organic matter $[M(COD)L^{-3}]$	Slowly biodegradable substrate $[M(COD)L^{-3}]$	Active heterotrophic biomass $[M(COD)L^{-3}]$	Active autotrophic biomass $[M(COD)L^{-3}]$	Particulate products arising from biomass decay $[M(COD)L^{-3}]$	Oxygen (negative COD) $[M(-COD)L^{-3}]$	Nitrate and nitrite nitrogen $[M(N)L^{-3}]$	$NH_4^+ + NH_3$ nitrogen $[M(N)L^{-3}]$	Soluble biodegradable organic nitrogen $[M(N)L^{-3}]$	Particulate biodegradable organic nitrogen $[M(N)L^{-3}]$	Alkalinity – Molar units	Kinetic Parameters: Heterotrophic growth and decay: $\hat{\mu}_H$, K_S, $K_{O,H}$, K_{NO}, b_H; Autotrophic growth and decay: $\hat{\mu}_A$, K_{NH}, $K_{O,A}$, b_A; Correction factor for anoxic growth of heterotrophs: η_g; Ammonification: k_a; Hydrolysis: k_h, K_X; Correction factor for anoxic hydrolysis: η_h

to maintain consistent units. Although not expressed explicitly in the model, nitrate nitrogen will also be removed by biomass decay. Like oxygen removal, this is accomplished by the recycling of organic matter during decay, making it available for anoxic growth of heterotrophic biomass.

The i = 10 column contains soluble ammonia nitrogen, S_{NH}, which is assumed to be the sum of the ionized (ammonium) and un-ionized (ammonia) forms. However, the un-ionized form is insignificant at pH values near neutrality so it is satisfactory simply to write models for 'ammonia' oxidation in terms of the total ammonium nitrogen concentration. Examination of the stoichiometric coefficients in the i = 10 column reveals that ammonia nitrogen is formed by ammonification of soluble biodegradable organic nitrogen and is removed by growth of the biomass. The major sink for the ammonia nitrogen is as the energy source for aerobic growth of the autotrophic biomass ($-1/Y_A$). However, nitrogen is also incorporated into biomass during cell synthesis and a term is included ($-i_{XB}$) for the nitrogen used during growth of both heterotrophs and autotrophs.

The i = 11 column contains the soluble organic nitrogen, S_{ND}, which is formed by hydrolysis of particulate organic nitrogen and converted to ammonia nitrogen by ammonification. Particulate biodegradable organic nitrogen, X_{ND}, is given in the i = 12 column. It is generated from decay of both heterotrophic and autotrophic biomass, i_{XB}, minus the amount associated with the inert particulate products, $f_P i_{XP}$, and is lost by ammonification. Although this organic nitrogen is particulate, it is not added to the other particulate forms to obtain the volatile solids concentration. This is because it is a subset of those materials and has already been included in their concentrations.

Three other forms of organic nitrogen will be present in the system: that associated with the biomass, X_{NB}; that associated with the particulate products, X_{NP}; and that associated with the inert particulate organic matter, X_{NI}. The concentration of each of these components can be calculated simply by multiplying X_B by i_{XB}, X_P by i_{XP}, and X_I by i_{XI}, the respective fractions of nitrogen present. These components are not currently needed in the matrix because the evolution of nitrogen gas (N_2) during denitrification is not included and thus a continuity check on nitrogen cannot be performed. Consequently, columns are not included for them. However, the model could easily be extended to incorporate nitrogen gas production, thereby allowing evaluation of potential problems in settling. If that were done, it would be necessary to include columns for these terms.

These 12 components discussed are considered to be the minimum required to model adequately an activated sludge system performing carbon oxidation, nitrification, and denitrification. Consequently, the complete model must include 12 mass balance equations. Of course, if a system is being designed to perform only one or two of these objectives, then appropriate components can be eliminated thereby decreasing the number of mass balance equations required in the model.

The i = 13 column represents total alkalinity, S_{ALK}. Incorporation of alkalinity into the model is not essential, but its inclusion is desirable because it provides information whereby undue changes in pH can be predicted. All reactions that involve the addition or removal of species with a proton accepting capacity and/or the addition or removal of protons will cause changes in alkalinity. Examples of the former are not included in the model because they are usually not significant in the activated sludge systems under consideration. Several examples of the latter are included in the model and are shown in Table 2. One is the conversion of ammonia nitrogen to amino acids during synthesis of heterotrophic and autotrophic biomass and the reversal of the process during ammonification (Scearce *et al.*, 1980). Another occurs during nitrification. When ammonium (NH_4^+) is oxidized for energy by autotrophs, eight electrons and ten protons are released; oxygen accepts eight electrons and eight protons so that there is a net release of two protons, thereby decreasing the alkalinity (Downing *et al.*, 1964). The last occurs during denitrification, because when nitrate (NO_3^-) acts as the electron acceptor, there is a net uptake of a proton, increasing alkalinity. Of the processes which add or remove protons, nitrification has the largest impact on alkalinity and can cause excessive decreases. From equilibrium chemistry of the carbonate system, if total alkalinity falls below about 50 g m^{-3} as calcium carbonate ($CaCO_3$) (1 mol total alk. m^{-3}), then the pH becomes unstable and can fall to values well below 6 (WRC, 1984). Low pH decreases the nitrification rate and causes other problems such as corrosive and aggressive effluents and bulking. Inclusion of the proper input term in a mass balance equation for alkalinity permits a user to evaluate whether the process configuration under consideration allows sufficient recovery of alkalinity during denitrification to maintain the pH in the proper range regardless of the proton release during nitrification. If not, then appropriate chemicals, such as lime, must be added to maintain the proper pH.

Processes in the Model

The fundamental processes incorporated into the model are listed in the leftmost column of Table 2, while their rate expressions are listed in the rightmost column. Basically, four processes are considered: growth of biomass, decay of biomass, ammonification of organic nitrogen, and 'hydrolysis' of particulate organics which are entrapped in the biofloc. To facilitate modelling, readily biodegradable material is considered to be the only substrate for growth of the heterotrophic biomass. Slowly biodegradable material is considered to be removed from suspension instantaneously by entrapment in the biofloc. Once there, it is acted upon by reactions which convert it into readily biodegradable substrate. These reactions are simply called 'hydrolysis' in the model, although in reality they are likely to be much more complex. The net result of their inclusion is to introduce a time delay into the utilization of oxygen since it is only associated with the growth of the organisms at the expense of readily biodegradable substrate. Decay is assumed to result in the transformation of active biomass into inert particulate products and into slowly biodegradable substrate which re-enters the cycle of hydrolysis, growth, etc. This allows more straightforward expression of decay under the various environmental conditions encountered in a single sludge system. It also has several important ramifications with respect to the values of the parameters, as will be discussed later.

First consider process 1, aerobic growth of heterotrophic biomass. Examination of row 1 in Table 2 shows that growth occurs at the expense of soluble substrate and results in the production of heterotrophic biomass. Associated with this is the utilization of oxygen. Since COD units are used for both substrate and biomass, and since oxygen may be considered to be negative COD, continuity requires that the oxygen requirement equal the net COD removal (soluble substrate removed minus cells formed). Ammonia nitrogen will be removed from solution and incorporated into cell mass. The kinetics of aerobic growth of the heterotrophic biomass are assumed to be subject to double nutrient limitation, with the concentrations of both readily biodegradable substrate and DO being rate determining. The effect of each constituent is modelled with a saturation function. It is recognized that a saturation function is not the ideal form for modelling substrate removal under dynamic conditions; however, the errors associated with its application to transients like those encountered in wastewater treatment systems are likely to be small. As discussed earlier, the primary purpose of the oxygen term is as a switching function which stops aerobic growth at low DO concentrations and thus the value of the saturation coefficient, $K_{O,H}$, is small. Removal of readily biodegradable substrate is considered to be proportional to growth. No provision is made for the storage of soluble substrate because that phenomenon is limited to only a few substrates such as soluble monosaccharides and acetate. However, it is widely recognized that substrates can be removed without associated biomass growth. This event is handled in the model through the immediate entrapment of slowly biodegradable substrate.

Row 2 represents anoxic growth of the heterotrophic biomass with nitrate nitrogen as the terminal electron acceptor. Like aerobic growth it occurs at the expense of readily biodegradable substrate and results in heterotrophic biomass. Nitrate nitrogen serves as the terminal electron acceptor and its removal is in proportion to the amount of readily biodegradable substrate removed minus the quantity of cells formed. As in aerobic growth, ammonia nitrogen is converted into organic nitrogen in the biomass. The rate expression for anoxic growth is analogous to the one for aerobic growth. In fact, the effect of readily biodegradable substrate on the rate is identical, including the value of the saturation coefficient, K_S. It is known, however, that the maximum rate of substrate removal under anoxic conditions is often less than it is under aerobic conditions. This could either be because $\hat{\mu}_H$ is lower under anoxic conditions or because only a fraction of the heterotrophic biomass is able to function with nitrate as the terminal electron acceptor. It is currently impossible to differentiate between these possibilities. Thus, from a modelling standpoint, the easiest way to incorporate the effect is to add an empirical coefficient, η_g, to the rate expression, where $\eta_g < 1.0$ (Batchelor, 1982). Anoxic growth depends upon the concentration of nitrate nitrogen in a manner analogous to the way in which aerobic growth depends upon the dissolved oxygen concentration. Furthermore, anoxic growth is inhibited when oxygen is present and the term $K_{O,H}/(K_{O,H}+S_O)$ is incorporated to reflect that fact. The coefficient $K_{O,H}$ has the same value as in the expression for aerobic growth so that as aerobic growth declines, anoxic growth increases. Like the other similar terms, its primary use is as a switching function.

Aerobic growth of autotrophic biomass is depicted in row 3 of Table 2. Soluble ammonia nitrogen serves as the energy source for growth of the nitrifiers resulting in autotrophic cell mass and nitrate nitrogen as end products. In addition, a small amount of ammonia is incorporated into the biomass. Oxygen is used in proportion to the amount of ammonia nitrogen oxidized. A double saturation function is used to express the dependency of the autotrophic specific growth rate upon the soluble concentrations of both ammonia nitrogen and oxygen, with the latter serving as a switching function. Both the saturation coefficients, K_{NH} and $K_{O,A}$, are small. Although aerobic growth of autotrophic biomass is known to be influenced by the pH of the wastewater in which the organisms are growing, this dependency was not included in the rate equation because of the difficulty of actually predicting the pH in a bioreactor. Rather, any potential problems with pH should be checked through use of the alkalinity term, as discussed earlier.

It is well established that the observed yield from the growth of heterotrophic biomass decreases as the SRT of a reactor is increased.

This phenomenon is thought to be due to many mechanisms, including predation, lysis, and the need for maintenance energy. Although it can be modelled in many ways, the most common technique under aerobic conditions is to incorporate all of the mechanisms into a single rate expression which is first order with respect to the concentration of active biomass and to let each unit of biomass COD lost result in the utilization of an equivalent amount of oxygen as done in the simple model in Table 1 (Grady and Lim, 1980). Even though this approach has worked well for the modelling of activated sludge systems performing only carbon oxidation and nitrification, many questions arise when the use of a terminal electron acceptor other than oxygen is considered. For example most studies suggest that decay continues under anoxic conditions, at least for the fraction of the biomass that can use nitrate nitrogen as the terminal electron acceptor. But what happens to the other heterotrophic biomass? Likewise, what happens to the denitrifying biomass when neither oxygen nor nitrate are present and anaerobic conditions prevail? It seems reasonable that, for many organisms, decay continues in a fermentative mode, but with no loss of COD because all organic oxidations would be coupled to organic reductions within the cell. All of this suggests that if decay were coupled directly to the utilization of the electron acceptor in the model, at least four separate rate expressions would be required: decay under aerobic conditions; decay under anoxic conditions of denitrifiers; decay under anoxic conditions of heterotrophic biomass incapable of denitrification; and decay under anaerobic conditions. Two problems arise with this approach. First, the equations would be complex, with a large number of switching functions. This would increase the complexity of the mass balance equations. Second, there are few fundamental data upon which to base the equations or with which to evaluate their parameters. Both of these suggest that a more pragmatic approach is warranted.

The approach adopted for modelling decay of the heterotrophic biomass is basically the death–regeneration concept of Dold *et al.* (1980), and is depicted in row 4 of Table 2. There it can be seen that the adopted rate expression is quite simple, i.e. first order with respect to the heterotrophic biomass concentration. The rate coefficient, however, is different in both concept and magnitude from the usual decay coefficient. In this case, decay acts to convert biomass to a combination of particulate products and slowly biodegradable substrate. No loss of COD is involved in this split and no electron acceptor is utilized. Furthermore, decay continues at a constant rate regardless of the environmental conditions (i.e. b_H is not a function of the type of electron acceptor or its concentration). The slowly biodegradable substrate formed is then hydrolysed, as depicted in row 7, releasing an equivalent amount of readily biodegradable COD. If conditions are aerobic, that substrate will be used to form new cells with concomitant oxygen uptake. If conditions are anoxic, cell growth will occur at the expense of nitrate nitrogen. If neither oxygen nor nitrate nitrogen are available, no conversion occurs and slowly biodegradable substrate will accumulate. Only when aerobic or anoxic conditions are resumed will it be converted and used.

The magnitude of the decay coefficient used herein will be different from that of the more usually encountered rate constant because of the recycling of substrate which occurs. In the usual technique, the loss of one unit of cell mass COD leads to the utilization of one unit of oxygen minus the COD of the inert particulate products formed. In this model, the loss of one unit of cell mass COD results in the ultimate formation of one unit of COD due to readily biodegradable substrate minus the COD of the inert particulate products formed. When the readily biodegradable COD is used for cell synthesis, only a fraction of a unit of oxygen will be required because of the energy incorporated into the cell mass. That cell mass must in turn undergo decay etc. before the unit of oxygen is finally removed. Consequently, to give the same amount of oxygen utilization per time due to decay, the decay coefficient must be larger. This has the result of increasing the turnover rate of cell mass, thereby making the actual microbial growth rate higher for a given solids retention time.

It should be emphasized that the modelling approach described above was adopted for pragmatic reasons. While the results from using such a model can mimic well the loss of biomass, consumption of electron acceptor etc. that occur in activated sludge systems (Dold *et al.*, 1980; Dold and Marais, 1986), there is no evidence that the model accurately reflects the actual mechanisms involved. It is obvious that the questions surrounding the effects of environmental conditions upon decay are badly in need of additional research.

The decay of autotrophs, given in row 5, is handled in exactly the same manner as the decay of heterotrophs. The justification for this is the likelihood that the decay observed in enrichment cultures of autotrophic bacteria is actually due to predation and lysis, with subsequent growth of adventitious heterotrophic bacteria upon the lysis products. While it is likely that the magnitude of the decay coefficient for autotrophic bacteria will be less than that for heterotrophic bacteria, even more questions can be raised about this process.

Another impact of biomass decay is to recycle nitrogen through the system. The conversion of biomass to slowly biodegradable substrate and then to readily biodegradable substrate has associated with it a parallel conversion of organic nitrogen to ammonia nitrogen. These reactions occur in the same way that biodegradable organic nitrogen from the feed is converted into ammonia nitrogen.

Soluble organic nitrogen is converted to ammonia nitrogen through the reaction depicted in row 6 of Table 2. This simple first order equation is empirical in nature but has been found to be adequate for modelling the conversion when coupled with the process rate equation for hydrolysis of entrapped organic nitrogen (Dold and Marais, 1986).

Rows 7 and 8 in Table 2 show the models that have been adopted for hydrolysis of slowly biodegradable organic matter and biodegradable organic nitrogen. The degradation of slowly biodegradable organic matter is very important to realistic modelling of activated sludge systems because it is primarily responsible for the attainment of realistic space–time and real time dependent electron acceptor profiles. Consequently, a great deal of effort was devoted to this topic by the task group. Within the past few years, the major changes and innovations in activated sludge modelling have been directed toward the development of equations depicting the fate of entrapped particulate or stored soluble substrates. Careful examination of all of the available literature revealed that very little experimental work has been conducted specifically on the kinetics and mechanisms of degradation of particulate organic material. Most studies in the wastewater treatment field have been done as part of complex model systems, thereby making it difficult to verify independently the portions dealing with hydrolysis and degradation of particulates. Nevertheless, it was evident that certain features were required in order for the overall system models to give realistic electron acceptor profiles. One was that the rate was first order with respect to the active heterotrophic biomass present. Another was that the rate appeared to saturate as the amount of entrapped substrate became large in proportion to the biomass. Finally, because of the need for enzyme synthesis it was reasoned that the rate would be dependent upon the concentration of electron acceptor present. Because nothing was known about 'hydrolysis' under anaerobic conditions except for the limited information on decay presented earlier, it was decided to assume that the rate goes to zero in the absence of both oxygen and nitrate. Examination of row 7 in Table 2 shows that all of these features were incorporated. The organic nitrogen was assumed to be uniformly distributed throughout the slowly biodegradable substrate so that the rate of hydrolysis of entrapped organic nitrogen would simply be proportional to the rate of hydrolysis of slowly biodegradable substrate.

Characterization of wastewater and estimation of parameter values

IN ORDER for a model to have utility in the design and operation of wastewater treatment systems, it must be possible to evaluate parameter values which are wastewater specific and to estimate concentrations of important components in the influent. Examination of Table 2 reveals that the model has 13 components and, with the exception of X_P, all of them may appear in the influent. In the recommended notation for the modelling of biological wastewater treatment systems (Grau *et al.*, 1982), numerical subscripts are used to denote the concentrations of components at specific locations. Assuming that some sort of pretreatment, such as sedimentation, precedes the biological treatment system, the subscript for influent concentrations to the biosystem would be 1. Consequently, the concentration of readily biodegradable substrate in the feed will be indicated as S_{S1}, the concentration of slowly biodegradable substrate as X_{S1}, etc. Table 2 also contains 19 parameters, five of which are stoichiometric. The other 14 are kinetic. Fortunately, some of these show little variation from waste to waste and may be considered to be constants. Because of the nature of the components it is necessary to characterize the influent in terms of them at the same time that the stoichiometric parameters are being evaluated. Consequently, we will first describe a technique for doing that and then we will summarize techniques for evaluating the kinetic parameters.

Characterization of wastewater and estimation of stoichiometric coefficients

The most important factor by which a model can be judged is its ability to predict real time and space–time dependent changes in the requirement for the electron acceptor. It was because of this that substrate was partitioned into two fractions: readily and slowly biodegradable. These are operationally defined fractions which do not necessarily correspond to readily distinguishable physical characteristics such as soluble and particulate. Consequently characterization of the influent must be accomplished experimentally in a way which ensures that the model can adequately predict the electron acceptor requirement. Another important factor during design is prediction of the activated sludge production rate because it determines the size of sludge handling facilities and the concentration of activated sludge associated with a given hydraulic retention time. The effect of net specific growth rate on the electron acceptor requirement and the sludge production rate can be determined most easily by operating steady state completely mixed activated sludge reactors in an aerobic mode at a number of SRTs. The data obtained can be used in concert with other tests to characterize the wastewater and evaluate the stoichiometric coefficients.

The total COD in the influent wastewater is made up of:

$$S_{S1}+X_{S1}+X_{I1}+S_{I1} \tag{8}$$

where:

S_{S1} is readily biodegradable substrate;
X_{S1} is slowly biodegradable substrate;
X_{I1} is inert suspended organic matter; and
S_{I1} is inert soluble organic matter.

The concentration of inert soluble organic matter may be determined easily. Simply remove an aliquot of the reactor contents from a completely mixed reactor treating the wastewater at an SRT in excess of 10 days and aerate it in a batch reactor. If samples are removed periodically and analysed for soluble COD, the concentration will either remain constant or will decrease with time. The former will occur if the concentration of readily biodegradable COD in the reactor is negligible whereas the latter will occur if it is not. The final residual soluble COD is the inert material, which is equal to the concentration in the feed, S_{I1}.

Before the concentration of readily biodegradable substrate can be obtained, the heterotrophic yield, Y_H, must be known. This can be estimated by observing the mass of cell material formed during removal of soluble substrate. An aliquot of wastewater should be settled and filtered to remove the particulate material. The filtrate, which contains only soluble organic matter, should be seeded lightly with acclimated biomass from one of the completely mixed reactors. Aliquots should be removed periodically and both the soluble COD and the total COD determined. The heterotrophic yield can be determined from:

$$\text{Cell COD} = \text{Total COD} - \text{Soluble COD} \tag{9}$$

$$Y_H = \frac{\Delta \text{ cell COD}}{\Delta \text{ soluble COD}} \tag{10}$$

If this is done several times, an approximate Y_H value may be determined. Any errors in this estimate will be compensated for in the determination of other parameters or influent concentrations.

Once Y_H is known, the concentration of readily biodegradable substrate in the influent, S_S, can be estimated by measuring the change in oxygen utilization rate (OUR) in a single completely mixed reactor operated at an SRT near 2 days under a daily cyclic square wave feeding pattern (12 h with feed; 12 h without feed) (Ekama *et al.*, 1986). As shown in Figure 1, there is a rapid drop in oxygen uptake rate following feed termination. This is because any accumulated readily biodegradable substrate is rapidly used. The OUR will not drop to zero, however, because the accumulated slowly biodegradable substrate will continue to be used at the same rate for a time period. Thus the immediate drop in OUR is associated only with the readily biodegradable material and can be used to find its concentration:

$$S_{S1} = \frac{\Delta OUR \times V}{Q(1 - Y_H)} \quad (11)$$

where:

ΔOUR is the change in OUR following feed termination ($ML^{-3}T^{-1}$);
V is the reactor volume (L^3);
Q is the feed flow rate prior to termination (L^3T^{-1}).

Having determined the concentrations in the wastewater of the total COD, readily biodegradable COD, and the inert soluble COD, it is only necessary to determine either the COD of the inert suspended organic matter, X_{I1}, or the COD of the slowly biodegradable substrate, X_{S1}, because the other can be determined by difference using Formula (8). It is recommended that the concentration of COD contributed by inert suspended organic material be evaluated as a parameter for fitting the model to data showing the effect of SRT on the sludge production.

The sludge in the activated sludge process comes from four major sources: growth of heterotrophic biomass on biodegradable substrate (S_S and X_S); production of inert particulate products by decay of the biomass; accumulation of inert suspended organic matter from the feed; and accumulation of undegraded slowly biodegradable substrate. Autotrophic biomass will also be present, but its contribution is so small for most wastewaters that it may be neglected in this analysis. Growth of heterotrophic biomass is proportional to the degradation of substrate with the proportionality constant being the heterotrophic yield, Y_H (Table 2). Decay of heterotrophic biomass occurs with a rate constant, b_H, and results in a fraction of the biomass, f_P, being transformed into inert particulate products. If mass balance equations are written which allow prediction of the effect of SRT on the sludge production

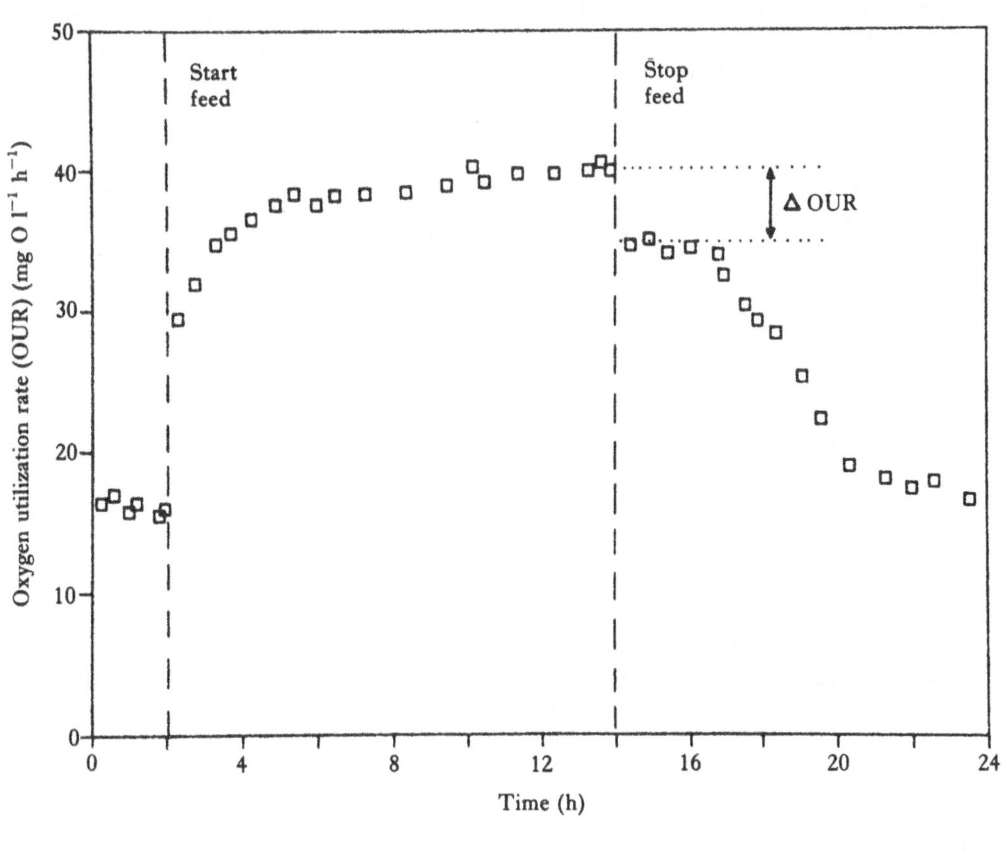

Fig. 1. Response of a completely mixed activated sludge reactor to a 12 h square wave response as used to determine the concentration of readily biodegradable substrate (Source: Ekama *et al.*, 1986).

rate in a completely mixed activated sludge system at steady state, it will be seen that the unknowns are X_{I1}, X_{S1}, f_P, b_H, S_S, k_h and K_X. The value of f_P is not likely to vary greatly from one wastewater to another because it is a characteristic of the biomass. For the type of model proposed here, in which decay results in the recycling of substrate, f_P has a value of 0.08 g COD (g COD)$^{-1}$ (Dold and Marais, 1986). The value of the decay rate constant, b_H, can be evaluated independently, as described later. If completely mixed activated sludge reactors are operated at steady state under conditions of constant mass and hydraulic loadings at SRTs in excess of 5 days the concentrations of readily, S_S, and slowly, X_S, biodegradable substrate in the reactor will generally be negligibly small compared to the amount in the feed (Ekama *et al.*, 1986). Recognition of this allows the hydrolysis parameters (k_h and K_X) and S_S to be dropped from the equations, thereby allowing them to be simplified. The concentration of slowly biodegradable substrate in the influent, X_{S1}, can be calculated from the total feed COD in terms of the inert particulate COD, X_{I1}, using Formula (8). Consequently, only X_{I1} needs to be evaluated since Y_H, f_P and b_H are independently known. This can be done by using a one dimensional search routine which chooses X_{I1} to minimize the error sum of squares when predicted sludge production rates are compared to measured rates as a function of SRT. This fitting acts to tune the model to the particular wastewater under study and compensates for any error made in Y_H and b_H during their estimation. Once X_{I1} is known, X_{S1} can be calculated from Formula (8). For variable strength influents, it can generally be assumed that the various fractions stay in constant proportion to one another.

In most activated sludge modelling, it is assumed that the concentration of biomass in the influent is negligible compared to the amount formed within the process. That approach is taken here, primarily because more research is needed regarding the impact of biomass in the influent. No procedure is recommended for measuring the influent concentrations. If there were a desire to include them in the model, appropriate microbiological methods would have to be employed.

Examination of Table 2 reveals that the model includes the soluble concentrations of oxygen, nitrate plus nitrite nitrogen, ammonia nitrogen, and alkalinity. The concentrations of all of these constituents in the feed may be measured by appropriate chemical tests.

Since the purpose of the model is to predict the performance of a single sludge system performing carbon oxidation, nitrification, and denitrification, it is important that the nitrogen be accounted for. Oxidizable nitrogen may be present in five forms: ammonia nitrogen, S_{NH}; soluble, inert organic nitrogen, S_{NI}; particulate, inert organic nitrogen, X_{NI}; readily biodegradable organic nitrogen, S_{ND}; and slowly biodegradable organic nitrogen, X_{ND}. As stated above, the concentration of ammonia nitrogen in the feed may be determined by appropriate analysis of a filtered sample. The concentration of soluble, inert organic nitrogen in the influent may be determined by performing Kjeldahl nitrogen tests on aliquots of the samples used to determine the soluble, inert COD. The Kjeldahl test may also be used to determine the total concentration of soluble organic nitrogen in the feed. Subtraction of the inert soluble organic nitrogen from that value approximates the readily biodegradable organic nitrogen, S_{NDI}. If the readily biodegradable and slowly biodegradable organic nitrogen in the feed are assumed to be proportioned in the same way as the readily biodegradable and slowly biodegradable COD in the feed, then the concentration of slowly biodegradable organic nitrogen in the feed may be determined from the concentration of readily biodegradable organic nitrogen in the feed:

$$\frac{S_{NDI}}{X_{NDI}+S_{NDI}}=\frac{S_{SI}}{X_{SI}+S_{SI}} \tag{12}$$

The only unknown is X_{NDI}, for which Equation (12) can be solved. There is no need to determine the particulate, inert nitrogen in the feed since nitrogen continuity cannot be checked because of the loss of nitrogen gas.

Three additional stoichiometric parameters must have values assigned to them. Because of the restricted nature of the nitrifying population in activated sludge, the autotrophic yield, Y_A, is not likely to vary much from system to system. Consequently, it should be adequate to use values obtained from the literature. An appropriate value appears to be 0.24 mg cell COD/mg N oxidized, which follows from the observation that 4.33 g of oxygen are used for each gram of nitrate nitrogen formed (Grady and Lim, 1980). The mass of nitrogen per mass of cell COD, i_{XB}, can be approximated closely enough by assuming that cell mass is represented by $C_5H_7O_2N$. The resultant value is 0.086 g N (g COD)$^{-1}$. The mass of nitrogen per mass of COD in the inert particulate products, i_{XP}, can also be approximated from literature values. An appropriate value is 0.06 g N (g COD)$^{-1}$.

Estimation of kinetic parameters

The purpose of the two half-saturation coefficients, $K_{O,H}$ and K_{NO}, is to serve as switching functions to shut off aerobic heterotrophic growth and start anoxic growth as the dissolved oxygen concentration drops. Likewise, the purpose of the oxygen half-saturation coefficient for the autotrophs, $K_{O,A}$, is to serve as a switching function stopping nitrification when the dissolved oxygen level gets too low. Consequently, the actual values used are not critical as long as they are of the appropriate order of magnitude and are small in comparison to operating concentrations. This suggests that it is not necessary to evaluate these parameters on a case by case basis. Rather the use of default values, to be given later, would be satisfactory.

The most critical parameter for characterizing the growth of the autotrophic biomass is $\hat{\mu}_A$, the maximum specific growth rate. This is because it is more sensitive to the constituents in the wastewater than is the half-saturation constant K_{NH} and because it determines the minimum SRT below which wash-out of the nitrifiers would occur. Consequently, every effort should be made to measure it accurately. The recommended procedure is to measure $\hat{\mu}_A$ during a dynamic test on one of the completely mixed reactors being run to determine the heterotrophic parameters, providing it is barely nitrifying and has a high DO concentration. By so doing, an accurate measure will be obtained of $\hat{\mu}_A$ in the actual wastewater environment. At the start of the test the sludge wastage rate from the completely mixed reactor is decreased to make the SRT greater than that required to achieve a high degree of nitrification. The concentration of nitrate nitrogen in the reactor should be measured over time as it increases through growth of additional nitrifying bacteria. Since the concentration of nitrate nitrogen is proportional to the mass of autotrophic bacteria in the sludge, the change in the nitrate concentration can be used to estimate $\hat{\mu}_A$ (Hall, 1974). If the natural logarithm of the nitrate nitrogen concentration is plotted versus time its slope will be $\hat{\mu}_A - 1/\theta_X - b'_A$ where θ_X is the new SRT and b'_A is the traditional decay rate coefficient for the nitrifiers. Since θ_X is known and b'_A may be assumed, $\hat{\mu}_A$ is known.

Unlike the situation for heterotrophic cell mass, the specific decay rate coefficient for autotrophic bacteria in this model, b_A, is numerically equivalent to the traditional decay rate constant, b'_A. This follows from the fact that the recycling of organic matter that results from decay occurs through the activity of the heterotrophic biomass and not the autotrophic biomass. There are a number of questions concerning the mechanisms by which autotrophic bacteria undergo decay. Consequently, there was general agreement among the task group members that it would be difficult to measure b_A with any real meaning. Examination of the research literature revealed that b_A should lie between 0.05 and 0.15 day^{-1} for most activated sludge conditions. Consequently, it is recommended that a value within that range be assumed.

The half-saturation coefficient for the nitrifying bacteria, K_{NH}, can be determined by the procedure of Williamson and McCarty (1975). Samples of nitrifying activated sludge from a completely mixed reactor are removed and placed into fed-batch reactors which receive continuous mass loadings of ammonia nitrogen below the maximum nitrification potential of the biomass. By using as feed wastewater spiked with additional ammonia nitrogen, it is possible to make the volumetric flow rate very small, thereby allowing each reactor to reach a pseudo-steady state. This provides information on the relationship between the specific nitrification rate and the pseudo-steady state ammonia nitrogen concentration which can be analysed by any of several techniques to provide a value for the half-saturation coefficient, K_{NH}.

Because of the influence of environmental factors such as pH, temperature and DO concentration on the rate of nitrification, special care should be taken in the preceding tests to maintain those factors at constant, appropriate values. It is especially important for the DO concentration to be maintained high enough to make the term $S_O/(K_{O,A}+S_O)$ approach unity.

The decay coefficient, b_H, is very important to predictions of sludge production and oxygen requirements, so it must be determined for the sludge in use. Sludge is removed from a completely mixed reactor and put into a batch reactor where the OUR can be measured many times over a period of several days (Ekama *et al.*, 1986). The slope of a plot of the natural logarithm of the oxygen uptake rate versus time will be the traditional decay coefficient, b'_H. Nitrification should be inhibited during the test by the addition of 20 mg l^{-1} of thiourea and the pH should be maintained at a constant value near neutrality. The model decay coefficient, b_H, can be calculated from:

$$b_H = \frac{b'_H}{1 - Y_H(1 - f_P)} \tag{13}$$

if Y_H and f_P are already known.

Two important parameters for the prediction of denitrification are η_g and η_h. The first is a correction factor which adjusts for either the change in $\hat{\mu}_H$ associated with anoxic conditions, or for the fact that only a portion of the biomass can denitrify. The second is a correction factor which adjusts for the observation that hydrolysis of slowly biodegradable organic matter occurs more slowly under anoxic conditions than under aerobic conditions. The two correction factors appear to have different numerical values, with η_h being the smaller of the two (Dold and Marais, 1986). Several factors are likely to influence the η values, including the fraction of bacteria in the influent that are capable of denitrification and the treatment system configuration, although theoretical calculations suggest that the latter is likely to be less important than the former (Henze, 1986). As a first approximation, η_g could be assumed to be equal to the ratio of the nitrate removal rate to the oxygen removal rate calculated on an oxygen equivalent basis using biomass harvested from the influent (Henze, 1986). However, after lab or pilot scale studies are under way, it will be possible to measure both η values directly using biomass from an experimental reactor.

The tests to measure η_g and η_h are performed at the same time by evaluating oxygen and nitrate consumption rates in two batch reactors which are equivalent in every respect except for the terminal electron acceptor (oxygen in one (aerobic) and nitrate in the other (anoxic)). The rationale for the tests is that immediately after bringing biomass into contact with wastewater in a batch reactor, the activity in the reactor will be dominated by growth of the heterotrophs on the readily biodegradable substrate whereas later activity will be predominantly due to use of substrate arising from hydrolysis of the slowly biodegradable substrate. When running the tests it is important that the ratio of substrate to biomass (F/M) be in the proper range as illustrated in Figure 2 (Ekama *et al.*, 1986). If F/M is too low the time during which the readily biodegradable substrate is removed will be too short to allow an accurate measurement of the OUR and nitrate utilization rate (NUR), whereas if it is too high the difference between the rates during the two phases will be too low to be clearly distinguishable. If the F/M is correct, the two zones of activity will be clearly distinguishable and of sufficient duration to allow accurate determination of the OUR in the aerobic reactor and the NUR in the anoxic reactor. If OUR_g represents the OUR during the first period and NUR_g represents the corresponding NUR, then:

$$\eta_g = \frac{2.86 \times NUR_g}{OUR_g} \tag{14}$$

Likewise, if OUR_h represents the OUR during the second period, and NUR_h the corresponding NUR, then:

$$\eta_h = \frac{2.86 \times NUR_h}{OUR_h} \tag{15}$$

The parameters describing biomass growth, $\hat{\mu}_H$ and K_S, are difficult to evaluate accurately, but that is not critical because the model is not very sensitive to their values. The main function of $\hat{\mu}_H$ is to allow the maximum OUR to be predicted. This suggests that measures of $\hat{\mu}_H$ should be based upon oxygen uptake measurements rather than cell growth or substrate removal. Because the concentration of readily biodegradable substrate in the effluent from an activated sludge system is generally quite low, it is not critical to the predictions of biomass concentration and OUR that it be modelled with high accuracy; i.e. an error factor of 2 or 3 will have little impact on model predictions. Consequently, the main function of K_S is as a switching function between first order and zero order kinetics for heterotrophic biomass growth and substrate removal. Cech *et al.* (1985) and Chudoba *et al.* (1985) have described a respirometric procedure for its measurement. Thus it seems appropriate to

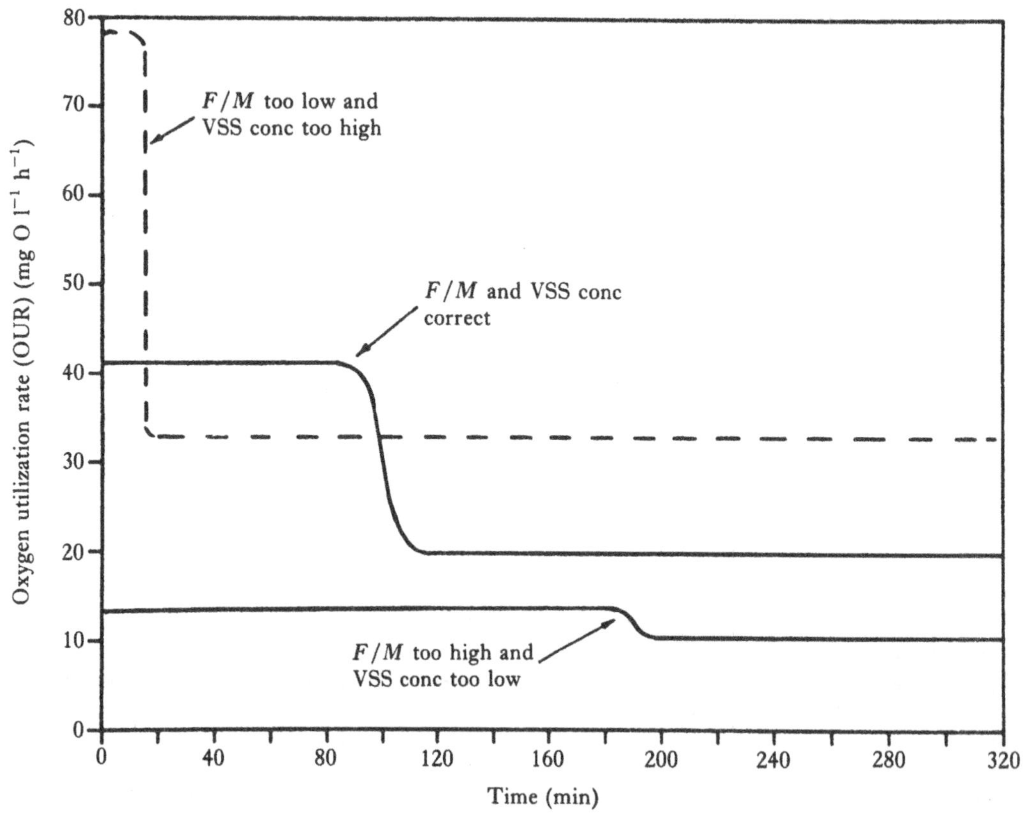

Fig. 2. Effect of changing the substrate to biomass ratio (food to micro-organism ratio, F/M) on the OUR in a batch reactor (Source: Ekama *et al.*, 1986). (VSS = volatile suspended solids.)

use respirometric techniques to estimate the values of both $\hat{\mu}_H$ and K_S.

In the procedure of Chudoba *et al.* (1985), biomass is removed from a laboratory-activated sludge reactor and aerated for 1 h to allow a constant background respiration rate to be achieved. Appropriate dilutions of the biomass are then contacted with dilutions of the wastewater chosen to allow various specific respiration rates up to the maximum to be achieved. During the test the DO concentration should be kept high so the term $S_O/(K_{O,H}+S_O)$ in the rate equation is made to approach unity. Use of the stoichiometric and kinetic parameters, in combination with the known characteristics of the wastewater, allow estimation of the heterotrophic biomass concentration, $X_{B,H}$, in the activated sludge from the lab-scale reactor. The measured respiration rates can then be divided by the heterotrophic biomass concentrations in the respirometer to obtain the specific respiration rates. Subtraction of the background rate from the measured rate gives the specific rate of substrate oxidation, $r_{resp,OX}$. The specific growth rate, μ_H, can then be calculated as:

$$\mu_H = \left(\frac{Y_H}{1-Y_H}\right) r_{resp,OX} \tag{16}$$

Because of the high DO concentration maintained during the test, μ_H is a function of only the readily biodegradable substrate concentration, S_S. The data on μ_H as a function of S_S may be analysed by any of several techniques to obtain $\hat{\mu}_H$ and K_S. This procedure is very sensitive to small changes in S_S and thus allows reasonable estimates of K_S, even when it is small.

An important factor which has only recently been recognized is that biomass grown in different reactor configurations exhibit different values of $\hat{\mu}_H$ and K_S even though the reactors are operated at the same SRT, loading, etc. (Cech *et al.*, 1985; Dold and Marais, 1986). Generally, the values of $\hat{\mu}_H$ and K_S tend to be lower for biomass grown in a completely mixed reactor with constant feed input than they are for biomass grown in a reactor which incorporates either time or space dependent changes in substrate concentration. Although the evidence is limited, this is probably due to predominance differences within the biomass brought on by different selective pressures in the two types of reactor. This suggests that care must be used in the collection and interpretation of kinetic data. Additional research is needed on this phenomenon, particularly with regard to the question of the best reactor configuration in which to grow biomass during parameter evaluation studies. In the meantime, preliminary evidence suggests that it would be acceptable to use a completely mixed reactor receiving a daily cyclic square wave input of feed. This is the same reactor configuration recommended for determination of the concentration of readily biodegradable substrate in the feed and thus it appears to be a useful tool for wastewater characterization studies.

The final parameters to be evaluated are the maximum specific hydrolysis rate k_h, the half-saturation coefficient, K_X, for hydrolysis of slowly biodegradable organic matter and the ammonification rate, k_a. Unlike $\hat{\mu}_H$ and K_S, these parameters appear to be relatively independent of the reactor configuration (Dold and Marais, 1986). In order to measure k_h, the biomass must be saturated with slowly biodegradable substrate. This, too, is most easily accomplished by operating a completely mixed activated sludge reactor at a short SRT with feed conforming to a daily cyclic square wave pattern (Ekama *et al.*, 1986). Figure 1 shows the pattern of oxygen uptake over the 24 h period and it will be recalled that the immediate drop upon feed cessation was used to determine the concentration of readily biodegradable substrate. In addition, the plateau in the OUR after feed cessation is due to degradation of organics released by hydrolysis of slowly biodegradable substrate. The existence of a sustained plateau is evidence that the biomass is saturated and that hydrolysis is occurring at the maximum rate, thereby allowing evaluation of k_h. Furthermore, the pattern at which the OUR falls off with time is determined by K_X. Consequently, the best way to estimate k_h and K_X is by curve-fitting techniques to match the response of the model to the oxygen uptake pattern in Fig. 1 (Dold and Marais, 1986). Since all other parameters have been selected, the only unknowns for the curve-fit are the two hydrolysis parameters, and the technique has been found to be quite sensitive to their values. A similar cyclic square wave feed experiment where nitrification is inhibited allows for determination of the

Table 3 **Parameters and characteristics which may be assumed**

Symbol	Name
Y_A	Yield for autotrophic biomass
b_A	Decay coefficient for autotrophic biomass
f_P	Fraction of biomass leading to particulate products
i_{XB}	Mass of nitrogen per mass of COD in biomass
i_{XP}	Mass of nitrogen per mass of COD in products from biomass
$K_{O,H}$	Oxygen half-saturation coefficient for heterotrophic biomass
K_{NO}	Nitrate half-saturation coefficient for denitrifying heterotrophic biomass
$K_{O,A}$	Oxygen half-saturation coefficient for autotrophic biomass

ammonification rate, based on the release of ammonia from soluble organic nitrogen during the non-feed period.

Two important points arise from the previous discussion. First, some parameters need not be measured because assumed values are satisfactory. These are summarized in Table 3. Second, evaluation of the remaining parameters as well as certain aspects of the waste characteristics must proceed in a particular order because the values of some are needed before others can be obtained. These are listed in Table 4 in the order of their determination. While every effort has been made to give the most reasonable procedure for evaluating parameters and wastewater characteristics, it should be recognized that some techniques are provisional. Better techniques are likely to be developed as more experience is gained in use of the model.

Table 4 **Parameters and characteristics which must be evaluated and information needed**

Symbol	Name	Prior information needed
S_{NOI}	Soluble nitrate nitrogen concentration in wastewater	
S_{NHI}	Soluble 'ammonia' nitrogen concentration in wastewater	
S_{II}	Soluble inert COD concentration in wastewater	
S_{NII}	Soluble inert organic nitrogen concentration in wastewater	
S_{NDI}	Soluble biodegradable organic nitrogen concentration in wastewater	S_{NII}
Y_H	Yield for heterotrophic biomass	
S_{SI}	Concentration of readily biodegradable COD in wastewater	Y_H
$\hat{\mu}_A$	Maximum specific growth rate for autotrophic biomass	b_A
K_{NH}	Ammonia half-saturation coefficient for autotrophic biomass	
b_H	Decay coefficient for heterotrophic biomass	Y_H, f_P
X_{II}	Inert suspended organic matter concentration in wastewater	f_P, b_H, S_{SI}, S_{II}
X_{SI}	Slowly biodegradable organic matter concentration in wastewater	X_{II}, S_{SI}, S_{II}
X_{NDI}	Slowly biodegradable organic nitrogen concentration in wastewater	S_{SI}, X_{SI}, S_{NDI}
η_g	Correction factor for μ_H under anoxic conditions	
η_h	Correction factor for hydrolysis under anoxic conditions	
$\hat{\mu}_H$	Maximum specific growth rate for heterotrophic biomass	$Y_H, X_{SI}, X_{II}, S_{SI}, f_P$
K_S	Half-saturation coefficient for heterotrophic biomass	$Y_H, X_{SI}, X_{II}, X_{SI}, f_P$
k_h	Maximum specific hydrolysis rate	
K_X	Half-saturation coefficient for hydrolysis of slowly biodegradable substrate	
k_a	Ammonification rate	

Typical parameter ranges, default values, and effects of environmental factors

Typical parameter values

CONSIDER first those parameters and characteristics whose values may be assumed rather than evaluated for each situation. They are listed in Table 3.

The autotrophic yield, Y_A, is a composite value for the combined growth of *Nitrosomonas* sp. and *Nitrobacter* sp. A range of values has been reported in the literature, although in this instance the range is more likely to have been the result of different environmental conditions than of different organisms with differing metabolic efficiencies. Reported values range from 0.07 to 0.28 g cell COD formed (g N oxidized)$^{-1}$. The theoretical value associated with the observation that 4.33 g of oxygen are required per gramme of nitrate nitrogen formed is 0.24 g cell COD formed (g N oxidized)$^{-1}$.

As discussed earlier, there was general agreement within the task group that it would be difficult to measure the specific decay coefficient for autotrophic biomass with any real meaning. Although values in the range between 0.05 and 0.15 day^{-1} have been reported, relatively little is actually known about its value.

The coefficient, f_P, represents the fraction of the biomass that ends up as inert particulate products following decay. Typically, about 20% of the biomass formed is considered to contribute to the inert residue and thus f_P is usually given a value of 0.20 in traditional models. It should be recognized, however, that in this model decay results in the recycling of biomass through the synthesis–resolubilization route. Thus, in order to have the observed fraction of inert products formed per net unit of mixed-liquor volatile suspended solids (MLVSS) equal to about 20%, the fraction of them actually formed during each passage around the cycle must be less than 20%. This follows from the fact that the observed fraction equals:

$$\frac{f_P}{1 - Y_H(1 - f_P)} \qquad (17)$$

If the observed fraction is 20%, then the value of f_P for this model should be around 0.08.

Two other stoichiometric coefficients whose values may be assumed are the mass of nitrogen per mass of cellular COD in the biomass and in the inert particulate products. For a typical cell formulation ($C_5H_7O_2N$), the value of i_{XB} would be 0.086 g N (g cell COD)$^{-1}$. It is likely that the inert particulate products will contain less nitrogen and thus an appropriate value for i_{XP} would be in the region of 0.06 g N (g COD)$^{-1}$.

As discussed earlier, the purpose of the half-saturation coefficients for the electron acceptors is to serve as switching functions to turn aerobic and anoxic growth on or off as the oxygen and nitrate concentrations vary. The half-saturation coefficient for dissolved oxygen, $K_{O,H}$, has not been well characterized but is known to vary considerably from organism to organism. For example Lau *et al.* (1984) reported a value of 0.15 g O_2 m^{-3} for a floc-forming bacterium but only 0.01 g O_2 m^{-3} for the filamentous bacterium *Sphaerotilus natans*. As a result of its importance to a complete description of denitrification kinetics, K_{NO} has received more study. All have found it to be quite low, so that for most purposes denitrification behaves in a zero order manner with respect to nitrate concentration. Typical values range from 0.1–0.2 g NO_3–N m^{-3}. The half-saturation coefficient for the effect of dissolved oxygen on the nitrifying bacteria is important for incorporating the retardant effect that low DO levels have been observed to have. Values reported in the literature have ranged from 0.5 to 2.0 g O_2 m^{-3}. Parker *et al.* (1975) used a value of 1.3 g O_2 m^{-3} for illustrative purposes.

The parameter values which must be evaluated for each wastewater are listed in Table 4. The order in which they must be evaluated is also indicated.

The heterotrophic yield, Y_H, depends upon the nature of the substrate as well as the population of micro-organisms carrying out the degradation. For various pure cultures growing on a number of single substrates, Y_H has been observed to range from 0.46 to 0.69 g cell COD formed (g substrate COD removed)$^{-1}$. Yield values for mixed cultures growing on multicomponent substrates have been found in the same range. If the influent wastewater contains appreciable quantities of micro-organisms which are not explicitly enumerated during characterization of the wastewater, their presence may influence the observed value of Y_H. To date, relatively little research has been done on their impact.

The parameter $\hat{\mu}_A$ is one of the more critical parameters in the model because it determines the SRT at which washout of the nitrifying bacteria occurs. Because nitrification is being modelled as a single step process and because

Nitrobacter spp. are generally considered to have a higher maximum specific growth rate than *Nitrosomonas* spp., it is appropriate to use the $\hat{\mu}_A$ value associated with the removal of ammonia nitrogen (i.e. with *Nitrosomonas* spp. growth) in the model. Values have been reported in the literature which range from 0.34 to 0.65 day^{-1} for mixed cultures oxidizing ammonia nitrogen under optimal conditions in the laboratory. Because nitrifying bacteria are influenced by many environmental factors, such as pH and temperature, it is important that actual values be measured for the particular waste in question.

The decay rate coefficient, b_H, is important because it has a large effect on the predicted cell mass at any given SRT. In traditional modelling, there is no recycling of substrate from decay, as there is in this model. Consequently, it is difficult to compare values of the traditional decay rates, b'_H, with values of the modified parameter, b_H, used herein. Reported values of b'_H vary widely, ranging from lows of 0.05 day^{-1} for domestic sewage in the USA to highs of 1.6 day^{-1} for some food-processing wastes. It is this wide range that led to the recommendation that the decay rate coefficient be measured for each wastewater treatment situation under consideration.

The denitrification correction factor, η_g, must be included to account for the fact that either the maximum rate of readily biodegradable substrate removal per unit of biomass is lower under anoxic conditions than under aerobic, or that not all heterotrophic bacteria can use nitrate as the terminal electron acceptor. Although relatively few measurements have been made of this parameter it appears to fall in the range 0.6–1.0. The lower value appears to be associated with wastewaters from anaerobic sewers whereas the higher value seems to be associated with wastewaters from aerobic sewers.

Two other parameters listed in Table 4 are $\hat{\mu}_H$ and K_S which describe the growth of heterotrophs on the readily degradable substrate. They are very dependent on the nature of the wastewater being treated and thus large ranges of values have been reported in the literature. Furthermore, as discussed previously, they appear to be influenced by the configuration of the reactor within which the biomass is grown. Consequently, even for domestic sewage, values of $\hat{\mu}_H$ have been reported which vary from 3.0 to 13.2 day^{-1} while K_S values from 10 to 180 g m^{-3} of biodegradable COD have been given.

The maximum specific hydrolysis rate, k_h, the half-saturation coefficient for hydrolysis of slowly biodegradable substrate, K_X, and the ammonification rate, k_a are relatively new parameters for which little information exists. Thus it is not possible at this time to give the ranges within which the values are likely to lie.

The parameter η_h acts to decrease the maximum hydrolysis rate under anoxic conditions. Although information on it is also limited, it appears to have a value in the region of 0.4 (Dold and Marais, 1986).

Default values

The parameter values used by the task group in the modelling studies reported here are listed in Table 5 for 10 °C and 20 °C. Likewise, typical wastewater characteristics are listed in

Table 5 Typical parameter values at neutral pH

Symbol	Unit	Value at 20 °C	Value at 10 °C
Stoichiometric parameters			
Y_A	g cell COD formed (g N oxidized)$^{-1}$	0.24	0.24
Y_H	g cell COD formed (g COD oxidized)$^{-1}$	0.67	0.67
f_P	dimensionless	0.08	0.08
i_{XB}	g N(g COD)$^{-1}$ in biomass	0.086	0.086
i_{XE}	g N(g COD)$^{-1}$ in endogenous mass	0.06	0.06
Kinetic parameters			
$\hat{\mu}_H$	day^{-1}	6.0	3.0
K_S	g COD m^{-3}	20.0	20.0
$K_{O,H}$	g O_2 m^{-3}	0.20	0.20
K_{NO}	g NO_3–N m^{-3}	0.50	0.50
b_H	day^{-1}	0.62	0.20
η_g	dimensionless	0.8	0.8
η_h		0.4	0.4
k_h	g slowly biodegradable COD (g cell COD · day)$^{-1}$	3.0	1.0
K_X	g slowly biodegradable COD (g cell COD)$^{-1}$	0.03	0.01
$\hat{\mu}_A$	day^{-1}	0.80	0.3
K_{NH}	g NH_3–N m^{-3}	1.0	1.0
$K_{O,A}$	g O_2 m^{-3}	0.4	0.4
k_a	m^3 · COD (g · day)$^{-1}$	0.08	0.04

Table 6 **Typical characteristics of settled domestic sewage**

Symbol	Unit	Denmark	Switzerland	Hungary
S_S	$g\,COD\,m^{-3}$	125	70	100
S_I	$g\,COD\,m^{-3}$	40	25	30
X_S	$g\,COD\,m^{-3}$	250	100	150
X_I	$g\,COD\,m^{-3}$	100	25	70
S_{ND}	$g\,N\,m^{-3}$	8	5	10
X_{ND}	$g\,N\,m^{-3}$	10	10	15
S_{NH}	$g\,NH_3\text{–}N\,m^{-3}$	30	10	30
S_{NI}	$g\,N\,m^{-3}$	2	2	3
S_{NO}	$g\,NO_3\text{–}N\,m^{-3}$	0.5	1	1

Table 6 for several countries. The values in Table 5 are considered to be 'typical' for neutral pH and domestic wastewater. It should be emphasized, however, that many parameter values are strongly influenced by environmental conditions, as discussed in the next section. Thus, although the values in the tables may be used as default values in the absence of specific data, the danger in doing so should be recognized.

Environmental effects

Although a number of environmental factors can influence the parameter values, three, in particular, deserve mention. These are specific factors in the wastewater, pH, and temperature.

Most parameter values can be influenced by specific compounds in the wastewater, which may act in either a stimulatory or an inhibitory manner. This is particularly true of those which describe nitrification. Because of the many factors which can potentially have an effect, it is difficult to generalize about them. The safest approach is to evaluate the parameters on the specific wastewater in question, which is why that procedure was recommended.

The effects of pH on nitrification have been well documented and equations have been proposed in the literature which incorporate them. The pH also influences the kinetics of heterotrophic growth, but fewer quantitative relationships have been developed. Most estimates of parameter values have been made at neutral pH and thus it is implicitly assumed in the model that the pH is near neutrality and relatively constant. Because both nitrification and denitrification involve changes in the hydrogen ion concentration they are likely to alter the pH if the buffering capacity of the wastewater is not sufficient. Since the major constituent contributing to buffering capacity is alkalinity, the model was structured so that alkalinity changes can be calculated. This allows the user to check to be sure that the assumption of nearly constant pH is not violated.

Within a narrow temperature range (psychrophilic, mesophilic, or thermophilic) an increase in temperature generally results in an increase in the value of a rate coefficient like $\hat{\mu}$, b, or k_h in a manner that can be described by a modified Arrhenius equation. Because half-saturation coefficients are not rate coefficients, but are parameters which influence the shape of a μ–S (or ammonia nitrogen, oxygen etc.) curve it is more difficult to generalize about the effects of temperature on them. Some increase, some decrease, and some are unchanged. The important point to recognize, however, is that all kinetic parameters are influenced by temperature. This suggests that their values should be determined at the temperature which will impose the most critical condition in the full scale facility. If that cannot be done, then a correction factor for temperature must be applied. Although a number of temperature correction factors have been reported in the literature, most have been developed for isolated processes. Since no single study has determined the effects of temperature on all of the processes incorporated into this model, the task group was reluctant to mix correction factors from several studies.

Assumptions, restrictions and constraints

WHEN A wastewater treatment system is to be modelled, a certain number of simplifications and assumptions must be made in order to make the model tractable. Some of these are associated with the physical system itself, whereas others concern the mathematical model. Often these simplifications and assumptions are implicit, which may cause the user to overlook them. When that happens there is a strong likelihood that they will be violated, which could destroy the utility of the results. To prevent that from happening the following sections explicitly enumerate the major assumptions, restrictions and constraints associated with the model and the physical system it was designed to simulate.

Assumptions and restrictions associated with the model

1. The system operates at constant temperature. Because many of the coefficients are functions of temperature, their functionality would have to be explicitly expressed in the rate expressions, ρ_j, in order for time-variant temperature fluctuations to be considered.
2. The pH is constant and near neutrality. As discussed earlier, although it is known that the pH influences many of the coefficients, few expressions are available for expressing that influence. Consequently, constant pH has been assumed. The inclusion of the alkalinity in the model allows the user to detect potential problems with pH control.
3. No consideration has been given to changes in the nature of the organic matter within any given fraction (e.g. the readily biodegradable organic matter). In other words, the coefficients in the rate expressions have been assumed to have constant values. It is still possible, however, for the concentration associated with any influent fraction to vary with time. Thus while variable input loadings can be handled, changes in waste character cannot.
4. The effects of limitations of nitrogen, phosphorus, and other inorganic nutrients on the removal of organic substrate and on cell growth have not been considered. It is well known that inadequate inorganic nutrients can lead to problems in sludge settleability. Thus, care must be taken to be sure that sufficient quantities of inorganic nutrients are present to allow balanced growth.
5. The correction factors for denitrification, η_g and η_h, are fixed and constant for a given wastewater. It is possible that their values may be influenced by system configuration but this is not considered.
6. The coefficients for nitrification are assumed to be constant and to incorporate any inhibitory effects that other waste constituents are likely to have on them.
7. The heterotrophic biomass is homogeneous and does not undergo changes in species diversity with time. This assumption is inherent in the assumption of constant kinetic parameters. It also means that the effects of substrate concentration gradients, reactor configuration, etc. on sludge settleability are not considered.
8. The entrapment of particulate organic matter in the biomass is assumed to be instantaneous.
9. Hydrolysis of organic matter and organic nitrogen are coupled and occur simultaneously with equal rates.
10. The type of electron acceptor present does not affect the loss of active biomass by decay.

Constraints upon the application of the model

The following represent some of the constraints which must not be violated if simulation results are to have practical utility. These are necessary because things which are possible mathematically may not be possible in the real world.

1. The net growth rate or SRT of the biomass must be within the range that allows a flocculent biomass to develop. For example if the SRT falls below 3 days, there are likely to be severe problems with sludge settleability in an activated sludge system. Since the model does not consider sludge settling, the user must ensure that all conditions employed will result in a sludge which

settles properly. The upper limit on SRT for validity of the model to single sludge systems is not well established, but appears to be about 30 days.

2. Proper sludge settling is also dependent upon the concentration of solids entering the final settler. Thus, while it is possible mathematically to make the reactor hydraulic retention time small by making the activated sludge concentration very large, such a trade-off may not work in practice because it may be difficult to get the highly concentrated sludge to settle sufficiently to obtain a clear effluent. Conversely, if the sludge concentration entering the settler is too low, a proper sludge blanket may not be established and a poor effluent may result. As a rough guideline, the activated sludge concentration (in COD units) should generally fall between 750 and 7 500 g m^{-3}, depending upon the type of pretreatment. If it does not, then the reactor sizes should be adjusted to increase or decrease it as needed.
3. The unaerated fraction of the reactor volume should not exceed 50% because sludge settling characteristics may deteriorate if it does.
4. The mixing intensity in an aerated reactor will be proportional to the power expended per unit volume for oxygen transfer. If that intensity exceeds 240 s^{-1}, excessive floc shear is likely to cause poor sludge settling. If the chosen reactor sizes cause the mixing intensity in any single reactor to be too great, then the design is not practical and new reactor sizes should be chosen. It should be noted that the choice of new reactor sizes will change the activated sludge concentration if the SRT is kept constant. Consequently, the constraints on mixing intensity and activated sludge concentration should be considered simultaneously.

Implementation of the activated sludge model

THE FULL potential of a complex scheme of conversion processes as presented for the activated sludge system in Table 2 can only be realized if equal care is devoted towards the physical system in which these processes will be active. The modelling of complex activated sludge reactor schemes under non-steady state conditions is the ultimate goal of this report. Mass balances, which relate the change of state of a system to transport and conversion processes, are the most convenient tool to model system performance. If activated sludge flow schemes are modelled as a combination of continuous stirred tank reactors (CSTRs), the mass balances may be written as a set of coupled ordinary differential equations, which for the kinetics introduced here (Table 2) are non-linear. Numerical integration techniques will usually be required to solve these equations.

The following sections are provided to help the less experienced process engineer develop software based on the proposed activated sludge kinetics. A personal computer with a BASIC compiler is sufficient to implement a powerful program. Full listings of computer programs were purposefully not provided in this report. By requiring each user to develop his own software the task group hoped to make each user fully aware of the details in the model.

Modelling of complex flow schemes

The general flow scheme indicated in Fig. 3 may be used to model a variety of continuous flow activated sludge processes. Each reactor compartment, k, is assumed to be completely mixed and must be defined with regard to:

Volume, $V(k)$
Influent flow rate, $Q_{tot} \cdot \text{fraction}\ (k) = Q(k)$
Value of the mass transfer coefficient, K_La, of the aeration equipment under the operating conditions imposed. Alternatively, the DO concentration in each reactor could be fixed at a specified value to be maintained by a DO controller which would alter the K_La value as needed.

Figure 4 presents some possibilities for modelling typical activated sludge flow schemes. The right-hand column of Fig. 4 indicates how these reactor schemes may be characterized in numerical terms.

Definition of initial conditions

The activated sludge model includes 13 independent state variables for each reactor compartment. This requires the definition of a large number of initial values before integration may start. Since good estimates for most of these initial values are not usually available, one technique would be to start the dynamic simulation from a steady state situation as derived for a mean load condition. The steady state may be obtained by relaxation over several SRTs of all components of interest; alternatively a more direct procedure may be used as discussed later.

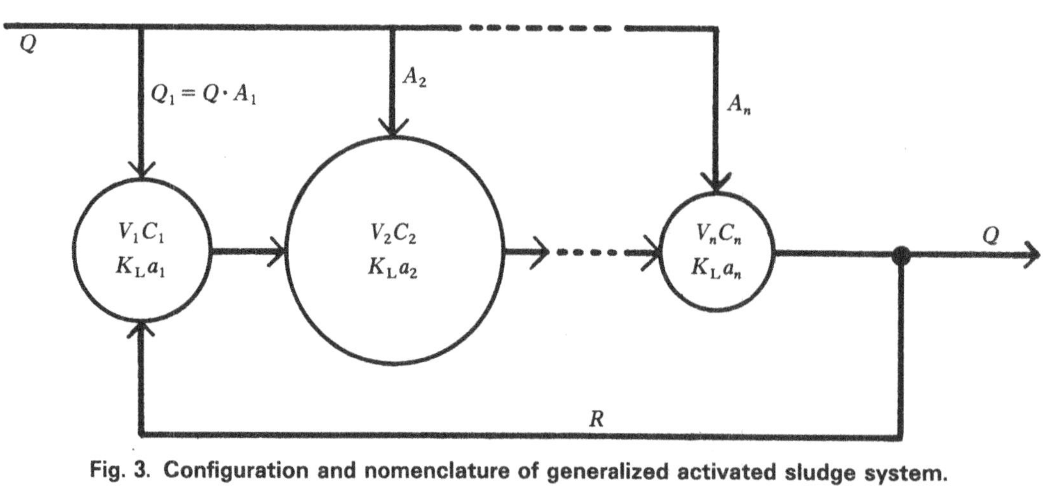

Fig. 3. Configuration and nomenclature of generalized activated sludge system.

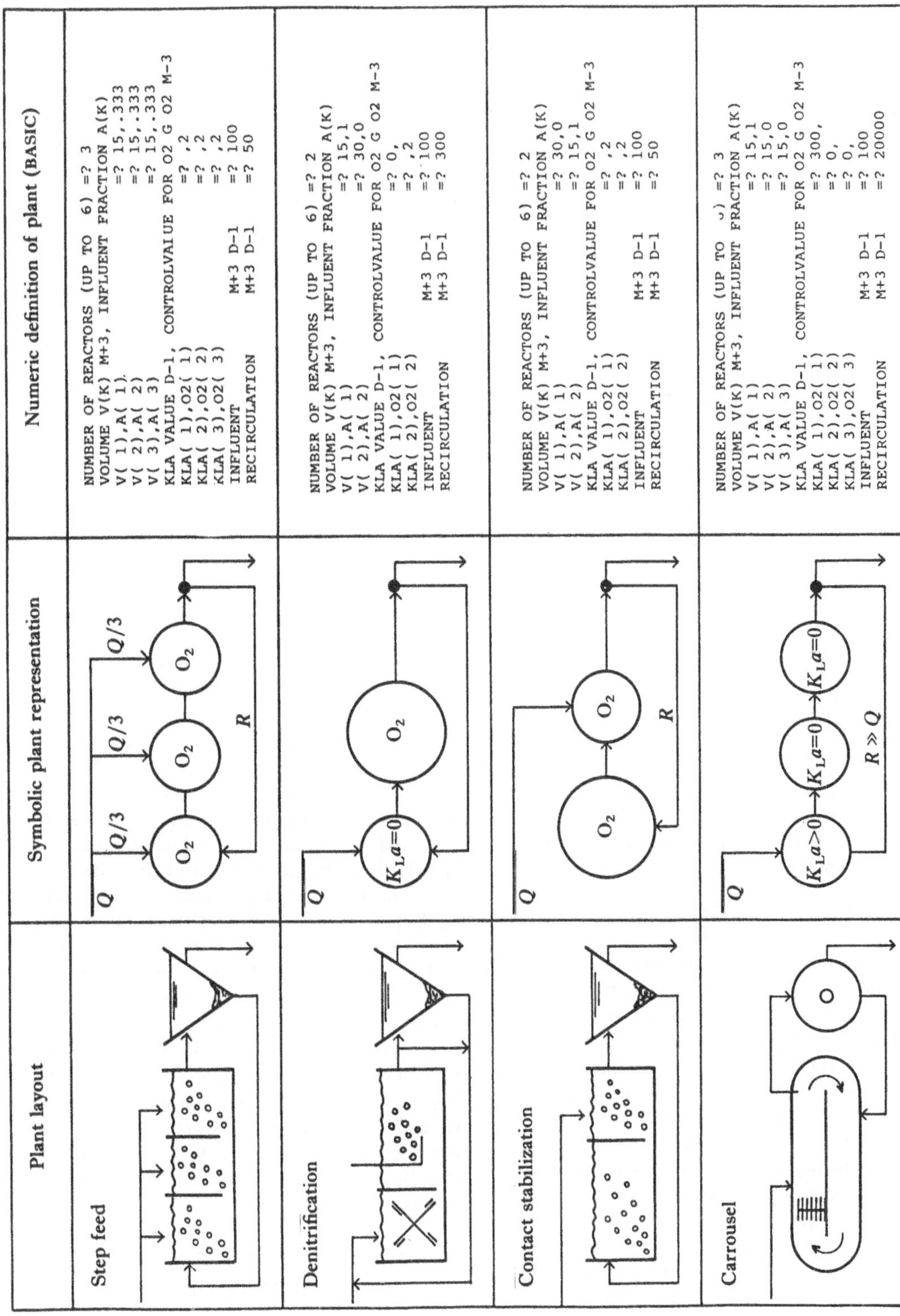

Fig. 4. Examples of systems which may be used to represent complex activated sludge flow schemes.

Consideration of variable load conditions

Diurnal or random variation of wastewater feed concentrations are typical for most treatment plants. For domestic wastewater treatment and for many industrial situations it is often sufficient to assume a constant input for a period of approximately 2 h and then to step to new values for the next 2 h, etc. However, with some process configurations, short time intervals are needed. The programmer should strive to choose the longest period consistent with reasonable results.

Factors governing step sizes during numerical integration

A commonly used method of numerical integration uses an equation of the form:

$$C(t+\Delta t) = C(t) + \left(\frac{dC}{dt}\right)\Delta t \tag{18}$$

where C represents a generalized state variable such as a concentration and Δt is the step size employed in the integration. The approach is most accurate when Δt is very small but the number of computations required (and hence the computer time to perform them) increases inversely with the size of Δt. Conversely, care must be taken not to make Δt too large, because to do so will result in large errors and other numerical problems. For example, if Δt is so large that $-(dC/dt)\,\Delta t > C(t)$, then $C(t+\Delta t)$ will be negative, which is physically impossible. Thus, one criterion for an upper limit on Δt is:

$$\Delta t < -C(t)\left(\frac{dC}{dt}\right)^{-1} \qquad (19)$$

For the generalized reactor system illustrated in Fig. 3, a mass balance for state variable i in reactor compartment k may be written as:

$$\frac{1}{C_{ki}}\frac{dC_{ki}}{dt} = \frac{F_{ki} - O_{ki} + P_{ki} - K_{ki}}{V_{ki}C_{ki}} \qquad (20)$$

where:

F and O are input (feed) and output transport terms (MT^{-1});
P and K are production and consumption terms (MT^{-1});
V is volume (L^3).

Combining Condition (19) and Equation (20), and neglecting the positive terms in the mass balance (F, P), results in an equation for the maximum step size:

$$\Delta t < \frac{V_k C_{ki}}{O_{ki} + K_{ki}} = \theta_{ki} \qquad (21)$$

The term θ_{ki} is the mean residence time of component i in reactor compartment k at steady state. The importance of Condition (21) is that it demonstrates that the maximum allowable step size for each component may be different, depending upon the mean residence time of that component. Recognition of this fact allows the numerical integration technique to be organized in a way which provides adequate accuracy for each component without wasting computational time.

To illustrate the importance of adjusting the step sizes for individual components, Condition (21) was used to calculate mean residence times using typical feed, hydraulic, stoichiometric, and kinetic parameters with reactor concentrations in the range which placed all saturation kinetic expressions in the first order region. This revealed that θ_{ki} was of the order of 10 min for $X_{B,H}$, $X_{B,A}$, X_P, X_S, and X_{ND}, whereas it was of the order of 1 min for S_S, S_{ND}, S_{NH}, and S_{ALK}. Furthermore, for S_O, θ_{ki} was of the order of 1 s. The fact that the mean residence times vary over a range of 10^3 means that much computational efficiency can be gained by using different step sizes for the various differential equations in the model. Consequently, the equations were partitioned into groups depending upon the step sizes appropriate for the components in them. For each group, the size of the step may be calculated based on Condition (19) with the following logic: for all components in the group and all reactor compartments, dC/dt is calculated and then a maximum value of:

$$\text{Abs}\,[C(t)(dC/dt)^{-1}] \qquad (22)$$

can be used to fix the size of the next time step. Sufficient accuracy is usually obtained when for each group the time step is chosen in the range of 5 to 20% of the above mentioned maximum value. This method of choice of the time step has the advantage that numerical problems do not usually appear, even though the time step is continuously kept at an upper limit.

A simple integration routine

Figure 5 illustrates an integration routine based upon Equation (18) and the partitioning of the differential equations according to the allowable step sizes. The routine integrates forward in time as shown.

The mean residence times, θ_{ki}, introduced in the preceding section are also useful for indicating the length of time required for each component to reach a pseudo-steady state given a fixed input. The relaxation time is directly proportional to θ_{ki}, and thus the oxygen concentration approaches steady state much faster than the soluble components, which reach steady state faster than the particulate components. In fact, the particulate components may take several SRTs to reach steady state. These facts may be used to develop an efficient routine for finding a steady state. The two internal loops (over Δt_2 and Δt_3) should be deactivated while maintaining Δt_1 and Δt_2. This would prevent the integration routine from integrating forward in real time, but would allow it to relax toward steady state more rapidly.

Structure of a possible program

Figure 6 indicates the structure of a possible simulation program. In co-ordination with the proposed integration routine, state variables should be grouped in a two dimensional array over reactor compartments and components such that one index brings together all components which are subject to equal integration steps (particulate, soluble, oxygen). This facilitates loop control. It is advantageous to introduce a stoichiometric matrix, ν_{ij} (according to Table 2) and to apply matrix algebra for the determination of the observed reaction rates, $r_{k,i}$, as a function of process rates $\rho_{k,j}$ and

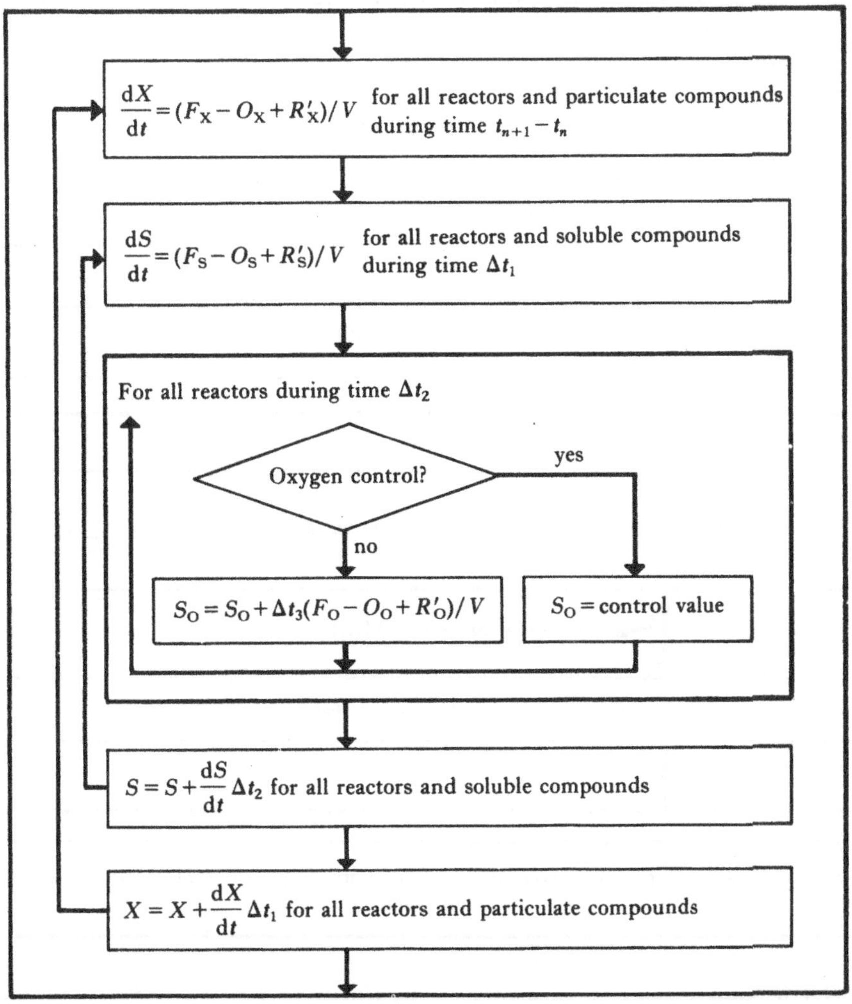

Fig. 5. Flow diagram of integration routine (*F* = input; *O* = output; *R'* = reaction).

stoichiometry. This approach allows for easy adjustment of process kinetic expressions and stoichiometry—a need which will invariably appear.

Examination of Figs 3 and 4 reveals that components are transferred from the last (n) reactor compartment to the first (1) by the recycle flow, R. This return of material must be handled in a manner which does not violate a mass balance on the secondary clarifier. The model as presented does not include any processes during clarification and thus the secondary clarifier is considered to be simply a separation point. A mass balance about that point for all soluble components, including oxygen reveals that the feed rate of a soluble component, i, from the last compartment to the first compartment is:

$$R \cdot S_{n,i} \tag{23}$$

The mass balance for particulate components must include sludge wastage and the inadvertent loss of solids in the overflow from the final clarifier. For the purpose of this illustration, the final clarifier has been assumed to be perfect (i.e. no particulates are lost). If sludge wastage is from the recycle line and if all particulate components settle together, then the feed rate of a particulate component, i, from the last compartment to the first compartment is:

$$(Q_{tot} + R)X_{n,i} - \left(\frac{\sum_{k=1}^{n} V_k \cdot X_{k,tot}}{SRT}\right)\left(\frac{X_{n,i}}{X_{k,tot}}\right) \tag{24}$$

Steady state solution for a single CSTR

As shown in Fig. 6, the initial values for use in the numerical integration routine can be obtained from the steady state solution for a single completely mixed reactor. This obviously requires simplification of the model because both nitrification and dentrification cannot occur simultaneously in a single reactor operated under constant conditions. The suggested approach is to set the DO concentration at a desired positive value so that both carbon oxidation and nitrification will occur. This, then, eliminates denitrification. Further simplifications may be made by assuming that all processes may be described by first order

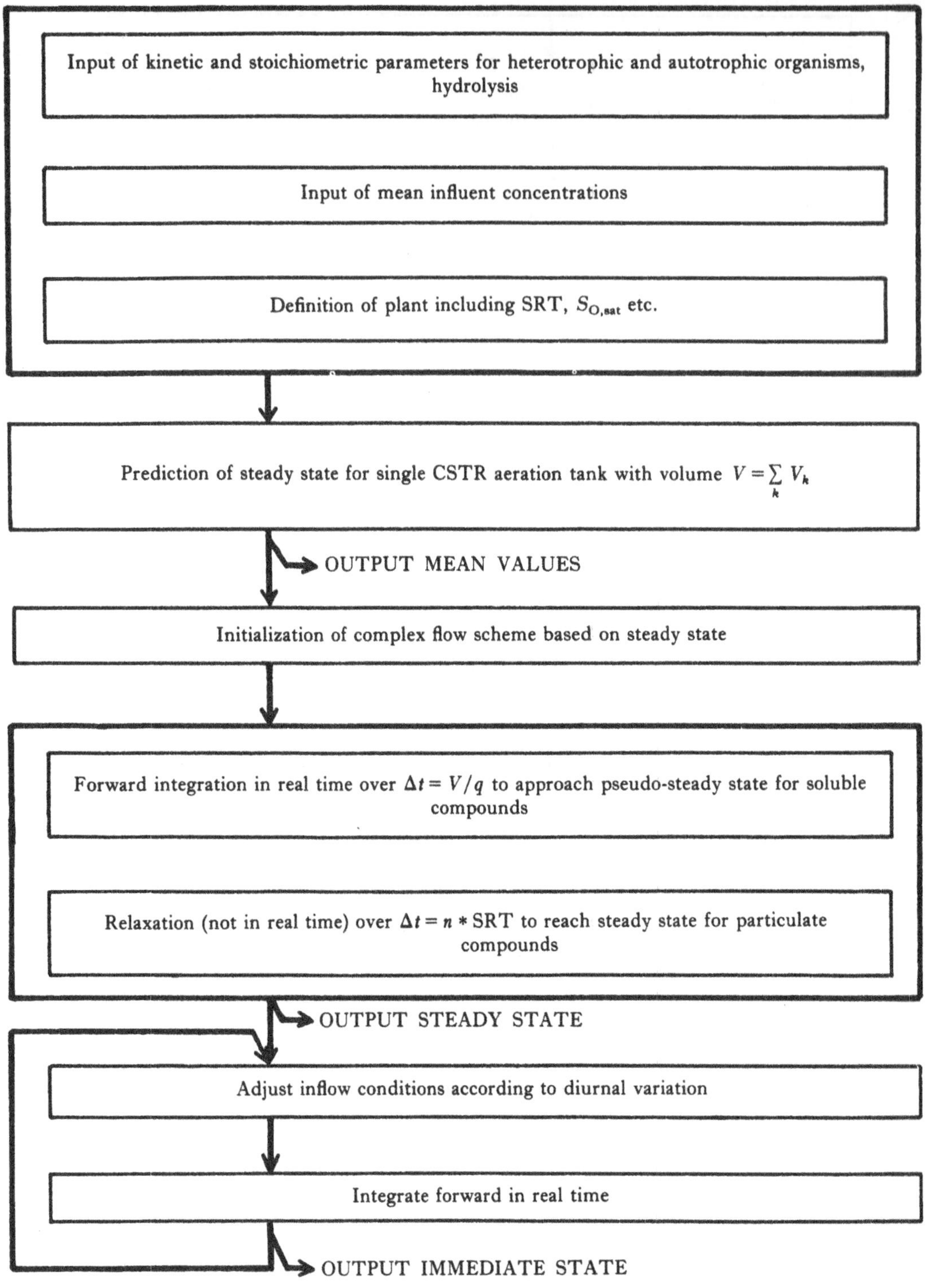

Fig. 6. Structure of a possible simulation program.

kinetics in the following form:

$$\rho_1 = k_1 S_1 \quad (25)$$

ρ_2 is not included (26)

$$\rho_3 = k_3 S_{NH} \quad (27)$$

$$\rho_4 = b_H X_{B,H} \quad (28)$$

$$\rho_5 = b_A X_{B,A} \quad (29)$$

$$\rho_6 = k_6 S_{ND} \quad (30)$$

$$\left.\begin{aligned}\rho_7 &= k_7 X_S \quad (31)\\ \rho_8 &= k_8 X_{ND} \quad (32)\end{aligned}\right\}\text{with } k_8 = k_7$$

With these linearizations, the mass balance for all components in a single completely mixed aeration tank yields the matrix shown in Table 7. The symbol, D, which is called the dilution rate, appears in the table for the first time. The dilution rate is simply the inverse of the residence time and two dilution rates may be defined as the hydraulic (soluble) dilution rate:

$$D_h = QV^{-1} \quad (33)$$

and the particulate dilution rate:

$$D_X = (\text{SRT})^{-1} \quad (34)$$

Table 7 Steady state solution for a single completely mixed aeration tank

	VECTOR OF STATE VARIABLES 1 S_I	2 S_S	3 X_S	4 X_S	5 $X_{B,H}$	6 $X_{B,A}$	7 X_P	8 K_La	9 S_{NO}	10 S_{NH}	11 S_{NO}	12 X_{ND}	13 S_{ALK}	FEED VECTOR
1	$-D_h$													$-D_hS_{I,1}$
2		$\nu_{21}K_1-D_h$		$\nu_{27}K_7$										$-D_hS_{S,1}$
3			$-D_X$											$-D_hX_{I,1}$
4				$\nu_{47}K_7-D_X$	$\nu_{44}b_H$	$\nu_{45}b_A$								$-D_hX_{S,1}$
5					$\nu_{54}b_H-D_X$									$-D_hX_{B,H,1}$
6						$\nu_{65}b_A-D_X$				$\nu_{63}K_3$				$-D_hX_{B,A,1}$
7					$\nu_{74}b_H$	$\nu_{75}b_A$	$-D_X$							$-D_hX_{P,1}$
8		$\nu_{81}K_1$						$O_{2sat}-S_O$		$\nu_{83}K_3$				$-D_h(S_{O,1}-S_O)$
9									$-D_h$	$\nu_{93}K_3$				$-D_hS_{NO,1}$
10		$\nu_{101}K_1$								$\nu_{103}K_3-D_h$	$\nu_{106}K_6$			$-D_hS_{NH,1}$
11											$\nu_{116}-D_h$	$\nu_{118}K_8$		$-D_hS_{NO,1}$
12					$\nu_{124}b_H$	$\nu_{125}b_A$						$\nu_{128}K_8-D_X$		$-D_hX_{ND,1}$
13		$\nu_{131}K_1$								$\nu_{133}K_3$	$\nu_{136}K_6$		$-D_h$	$-D_hS_{ALK,1}$

The second row reads
$\nu_{21}K_1S_S-D_hS_S+\nu_{27}K_7X_S=-D_hS_{S,1}$

Since the oxygen concentration in the aeration tank must be predefined with this approach, the required $K_L a$ value of the aeration equipment becomes the independent variable with subscript 8. Initial values for all state variables may now be obtained by matrix inversion and solution of the equations in Table 7. Such a solution may be obtained independently of the composition of the feed. This is especially valuable if aerobic stabilization of the sludge is to be described. The matrix in Table 7 can be set up very easily in a program, since all empty elements have corresponding ν_{ij} values of zero. Column 2 would as an example require the following BASIC statements for set-up:

```
For I = 1 to 13: MATRIX (2, 1)
        = NUE (2, 1)*K1: NEXT I
MATRIX (2, 2) = MATRIX (2, 2) − DH
```

Even if the stoichiometric matrix, ν, is changed, this procedure would still find valuable initial conditions. The activated sludge composition may then be obtained by matrix inversion of Table 7. This has the advantage that a change in stoichiometry results only in a change in the numerical values in the matrix to be inverted. No major changes in software are required.

If the assumption of first order kinetics results in initial estimates with so much error that the numerical integration is hampered, then it will be necessary to use saturation kinetics. The easiest way to do this is to leave the matrix as shown in Table 7, but to recognize that:

$$K_j = \frac{P_j}{C_j^*} \tag{35}$$

where C_j^* is the limiting compound for process j. This means that the matrix must be solved iteratively. First, a reasonable estimate is made of K_j and the equations are solved by matrix inversion. The resulting state variables are then used to obtain a new estimate of K_j and the process is repeated until it converges.

Sample output

Table 8 contains the sample output from a program of the model. The activated sludge system contains three reactors of equal size in series with the feed divided evenly between the first two. No oxygen is supplied to the first

Table 8 Sample output from activated sludge model

```
STOICHIOMETRIC PARAMETERS
=========================

HETEROTROPH. YIELD G COD G-1 COD=   0.570
FRACTION INERT     G COD G-1 COD=   0.100
N IN BIOMASS       G N G-1 COD  =   0.070
N IN INERT         G N G-1 COD  =   0.060
AUTOTROPHIC YIELD  G COD G-1 N  =   0.240

KINETIC PARAMETERS
==================

HETEROTROPHIC ORGANISMS

MUE MAX            D-1          =   4.000
KS COD             G COD M-3    =  20.000
KS O2              G O2 M-3     =   0.250
B DECAY            D-1          =   0.500
ETA DENITRIFICATION GROWTH      =   0.800
KS NO3-            G N M-3      =   0.500
HYDROLYSIS RATE    D-1          =   1.500
HYDROL SATUR RATIO G COD G-1 COD=   0.020
ETA DENITRIFICATION HYDROLYSIS  =   0.350
AMMONIFICAT RATE M+3 G-1 COD D-1=   0.080

AUTOTROPHIC ORGANISMS

MUE MAX AUT        D-1          =   0.879
KS NH4+            G N M-3      =   0.741
KS O2              G O2 M-3     =   0.500
KS HCO3-           MOL M-3      =   0.250
B ENDOGENOUS       D-1          =   0.132

INLET CONCENTRATIONS
====================

TOTAL   COD        G COD M-3    =          ? 400.0
SOLUBLE INERT COD  G COD M-3    =          ? 40.0
PARTICL INERT COD  G COD M-3    =          ? 40.0
KJELDAHL N TOTAL   G N M-3      =          ? 50.0
KJELDAHL N SOLUBLE G N M-3      =          ? 25.0
NO3 - N            G N M-3      =          ? 1.0
HCO3-              MOL M-3      =          ? 6.0

INLET CONCENTRATIONS ARE:
X INERT FEED       G COD M-3    =    40.000
X HETEROTROP       G COD M-3    =    96.000
X AUTOTROPHI       G COD M-3    =     0.001
X INE. DECAY       G COD M-3    =     0.000
X CODPARTICU       G COD M-3    =   160.000
X ORG N PART       G N M-3      =    18.280
S CODSOLUBLE       G COD M-3    =    64.000
S NH4-N            G N M-3      =    12.500
S NO3-N            G N M-3      =     1.000
S HCO3-            MOL HCO3 M   =     6.000
S CODINERT         G COD M-3    =    40.000
S ORG N SOL        G N M-3      =    10.100
S OXYGEN           G O2 M-3     =     0.000
```

```
DEFINITION OF PLANT
===================

NUMBER OF REACTORS (UP TO 10)   =         ? 3
VOLUME V(K) M+3, INFLUENT FRACTION ALFA(K)
V( 1),ALFA( 1)                  =         ? 15,0.5
V( 2),ALFA( 2)                  =         ? 15,0.5
V( 3),ALFA( 3)                  =         ? 15,0.0
KLA VALUE  D-1, CONTROLVALUE FOR O2  G O2 M-3
KLA( 1),O2CONTROL( 1)           =         ? 0,0
KLA( 2),O2CONTROL( 2)           =         ? 0,2
KLA( 3),O2CONTROL( 3)           =         ? 150,0
INFLUENT          M+3 D-1       =         ? 100
RECIRCULATION     M+3 D-1       =         ? 200
SRT (TOTAL)       D             =         ? 10
O2 SATURATION     G O2 M-3      =    8.637 (400 M MSL AT 20 DEG C)

INITIAL CONDITIONS ARE:    SINGLE CSTR AERATION TANK
=======================

X INERT FEED     888.89 G COD M-3
X HETEROTROP    1450.31 G COD M-3
X AUTOTROPHI      90.39 G COD M-3
X INE. DECAY     737.08 G COD M-3
X CODPARTICU      29.46 G COD M-3
X ORG N PART       2.54 G N M-3
S CODSOLUBLE       2.62 G COD M-3
S NH4-N            0.41 G N M-3
S NO3-N           33.31 G N M-3
S HCO3-            2.83 MOL HCO3 M-3
S CODINERT        40.00 G COD M-3
S ORG N SOL        0.93 G N M-3

X TOTAL         3196.12 G COD M-3
OBSERVED YIELD           = 0.4024 G COD G-1 COD
REQUIRED KLA VALUE       =  124.9 D-1
OXYGEN REQUIREMENT       =  829.2 G O3 M-3 D-1 TO MAINTAIN 2 G O2 M-3
EXCESS SLUDGE PRODUCTION=  143.8 G COD M-3 INFLUENT

RESULT OF RELAXATION OVER  10.0 DAYS
====================================

COMPOUND          INPUT REACTOR
                          1       2       3
X INERT FEED       40.0   999.7   835.6   831.4  G COD M-3
X HETEROTROP       96.0  1615.1  1363.3  1354.7  G COD M-3
X AUTOTROPHI        0.0   100.7    85.0    85.0  G COD M-3
X INE. DECAY        0.0   826.2   688.6   688.6  G COD M-3
X CODPARTICU      160.0    82.7    60.9    36.4  G COD M-3
X ORG N PART       18.3     7.2     5.4     3.0  G N M-3
S CODSOLUBLE       64.0     2.1     3.8     2.7  G COD M-3
S NH4-N            12.5     5.7     2.0     0.4  G N M-3
S NO3-N             1.0     7.8    14.3    18.0  G N M-3
S HCO3-             6.0     5.0     4.3     3.9  MOL HCO3 M-3
S CODINERT         40.0    40.0    40.0    40.0  G COD M-3
S ORG N SOL        10.1     0.7     1.2     0.9  G N M-3
S OXYGEN            0.0     0.0     2.0     3.0  G O2 M-3

X TOTAL                  3624.3  3033.3  2996.1  G COD M-3
O2 REQ HETEROTROPHIC       16.1   581.1   454.8  G O2 M-3 D-1
O2 REQ AUTOTROPHIC         23.5   743.1   365.4  G O2 M-3 D-1
O2 REQ TOTAL               39.6  1324.1   820.3  G O2 M-3 D-1
DENITRIFICATION RATE      119.0    19.6    10.3  G N M-3 D-1
```

reactor so denitrification occurs there. The oxygen concentration is fixed at 2.0 g m^{-3} in the second reactor and is supplied to the third at a fixed rate. The influent contains 160 g m^{-3} of slowly biodegradable substrate and 64 g m^{-3} of readily biodegradable substrate as well as the other constituents listed. The initial conditions calculated by matrix inversion are shown in the right column while the steady state output is shown in the lower portion of the same column. Examination of Table 8 will reveal that an engineer could rapidly evaluate several alternative reactor schemes using the program.

Conclusion

WHEN ONE considers a system as complex as a single-stage activated sludge system capable of carbon oxidation, nitrification, and denitrification, it is apparent that a tremendous investment of time and money would be required to operate a pilot plant at all of the possible conditions which might be considered during design. That means that our experience will always be limited. The availability of a model like the one presented here, however, in which rate equations are presented for the processes involved, allows the engineer to explore, through simulation, a very broad range of system configurations, inputs, and operational strategies. By so doing, his base of experience is greatly expanded and his intuitive decision-making ability is increased. Engineering design has always depended upon heuristic rules founded upon experience. By increasing their experiential base, the validity of those rules will be strengthened and the engineer's ability will be improved. Already, through the use of such models, it has been possible to develop general design guidelines for single sludge systems which give the engineer guidance about such factors as the maximum allowable TKN/COD ratios for complete denitrification, the maximum economic internal recycle ratio, and the maximum anoxic fraction in the reactor (Water Research Commission, 1984). Through continued application of the models it will be possible to define the feasible design space better, thereby reducing the alternatives which must be considered by a designer.

Once the parameter values have been calibrated to a particular wastewater, a model may be used by the engineer to eliminate inefficient designs and to choose those alternative system configurations which are most likely to be economic. For a given system flowsheet, there is more than one choice of unit sizes which will result in a desired degree of treatment. One job of the engineer is to choose those sizes which will do the desired job at least cost. Once a group of feasible designs has been separated from the other, less economic designs, the engineer must choose between them using suitable decision criteria. The availability of a usable mathematical model makes it possible to test a large number of potential designs in an economic manner, thereby ensuring that those chosen for inclusion in the final group are indeed sound.

After a plant has been built, a model like the one presented here can be used to evaluate the impact of new waste loads and to try new operational strategies. If management wishes to consider adding new discharges to the plant influent, the model may be used to predict their impact upon plant performance and to evaluate alternative operational strategies to mitigate that impact. If a plant is not performing as well as expected, alternative operational conditions can be tried with the model to see which are most likely to have a positive effect. In addition, alternative operational strategies could be tried to see which gave the greatest energy savings, which produced the least sludge, etc. In other words, the model again allows the engineer to expand his experience base without risk to the plant itself.

Finally, it should be recognized that modelling is an essential part of research which expands our knowledge base. The very exercise of creating a model requires the modeller to ask critical questions about the system being modelled. Often, as was the case here, the answers to those questions are less than satisfactory. This, then, pinpoints the need for more research, which will in turn lead to a new generation model based upon sounder principles. Hopefully, the model presented here will have that impact while also stimulating greater use of simulation by the engineering profession.

References

Batchelor, B. (1982). Kinetic Analysis of alternative configurations for single-sludge nitrification/denitrification. *J. Wat. Pollut. Control Fed., 54,* 1493–1504.

Cech, J. S., Chudoba, J. and Grau, P. (1985). Determination of kinetic constants of activated sludge microorganisms. *Wat. Sci. Technol., 17(2/3),* 259–272.

Chudoba, J., Cech, J. S., Farkac, J. and Grau, P. (1985). Control of activated sludge filamentous bulking; Experimental verification of a kinetic selection theory. *Wat. Res., 19,* 191–196.

Dold, P. L. and Marais, G. v. R. (1986). Evaluation of the general activated sludge model proposed by the IAWPRC task group. *Wat. Sci. Technol., 18(6),* 63–89.

Dold, P. L., Ekama, G. A. and Marais, G. v. R. (1980). A general model for the activated sludge process. *Prog. Wat. Technol., 12,* 47–77.

Downing, A. L., Painter, H. A. and Knowles, G. (1964). Nitrification in the activated sludge process. *J. Proc. Inst. Sew. Purif., 64,* 130–158.

Ekama, G. A., Dold, P. L. and Marais, G. v. R. (1986). Procedures for determining influent COD fractions and the maximum specific growth rate of heterotrophs in activated sludge systems. *Wat. Sci. Technol., 18(6),* 91–114.

Gaudy, A. F. Jr and Gaudy, E. T. (1971). Biological concepts for design and operation of the activated sludge process. *US Environmental Protection Agency Water Pollution Research Series, Report No. 17090, FQJ, 09/71.* US EPA, Washington, DC.

Grady, C. P. L. Jr and Lim, H. C. (1980). *Biological Wastewater Treatment, Theory and Applications.* Marcel Dekker, New York.

Grau, P., Sutton, P. M., Henze, M., Elmaleh, S., Grady, C. P. L. Jr, Gujer, W. and Koller, J. (1982). Recommended notation for use in the description of biological wastewater treatment processes. *Wat. Res., 16,* 1501–1505.

Hall, I. R. (1974). Some studies on nitrification in the activated sludge process. *Wat. Pollut. Control, 73,* 538–547.

Henze, M. (1986). Nitrate versus oxygen utilization rates in wastewater and activated sludge systems. *Wat. Sci. Technol., 18(6),* 115–122.

Herbert, D. (1958). Some principles of continuous culture. In: *Recent Progress in Microbiology,* G. Tunevall (Ed.). Almquist and Wiksell, Stockholm, pp 381–396.

Kountz, R. R. and Forney, C. Jr (1959). Metabolic energy balances in a total oxidation activated sludge system. *J. Wat. Pollut. Control Fed., 31,* 819–826.

Lau, A. D., Strom, P. F. and Jenkins, D. (1984). Growth kinetics of *Sphaerotilus natans* and a floc former in pure and dual continuous culture. *J. Wat. Pollut. Control Fed., 56,* 41–51.

McKinney, R. E. and Ooten, R. J. (1969). Concepts of complete mixing activated sludge. *Trans. 19th Sanit. Eng. Conf., University of Kansas,* pp 32–59.

Obayashi A. W. and Gaudy, A. F. Jr (1973). Aerobic digestion of extracellular microbial polysaccharides. *J. Wat. Pollut. Control Fed., 45,* 1584–1594.

Parker, D. S., Stone, R. W., Stenquist, R. J. and Culp, G. (1975). *Process Design Manual for Nitrogen Control.* US Environmental Protection Agency, Washington, DC. Technology Transfer.

Petersen, E. E. (1965). *Chemical Reaction Analysis.* Prentice-Hall, Englewood Cliffs, NJ.

Scearce, S. N., Benninger, R. W., Weber, A. S. and Sherrard, Y. H. (1980). Predictions of alkalinity changes in the activated sludge process. *J. Wat. Pollut. Control Fed., 52,* 399–405.

Van Haandel, A. C., Ekama, G. A. and Marais, G. v. R. (1981). The activated sludge process. Part 3: Single sludge denitrification. *Wat. Res., 15,* 1135–1152.

Water Research Commission (1984). *Theory, Design and Operation of Nutrient Removal Activated Sludge Processes.* Water Research Commission, PO Box 824, Pretoria 0001, South Africa.

Weddle, C. L. and Jenkins, D. (1971). The viability and activity of activated sludge. *Wat. Res., 5,* 621–640.

Williamson, K. J. and McCarty, P. L. (1975). Rapid measurement of Monod half-velocity coefficients for bacterial kinetics. *Biotechnol. Bioeng., 17,* 915–924.

ACTIVATED SLUDGE MODEL NO. 2

by

IAWQ TASK GROUP ON MATHEMATICAL MODELLING FOR DESIGN AND OPERATION OF BIOLOGICAL WASTEWATER TREATMENT

Mogens Henze, *Technical University of Denmark, Denmark*
Willi Gujer, *Swiss Federal Institute of Technology, Switzerland*
Takashi Mino, *Tokyo University, Japan*
Tomonori Matsuo, *Tokyo University, Japan*
Mark C. Wentzel, *Cape Town University, South Africa*
Gerrit v. R. Marais, *Cape Town University, South Africa*

Contents

1. Introduction

This report presents the results of the work by the IAWQ Task Group on *Mathematical Modelling for Design and Operation of Biological Wastewater Treatment Processes.* The **Activated Sludge Model No. 2 (ASM2)** presents a concept for dynamic simulation of combined biological processes for chemical oxygen demand (COD), nitrogen and phosphorus removal.

The model as presented here is a tool for:

- Research (testing results, selecting and optimizing experiments)
- Process optimization and troubleshooting at full-scale treatment plants
- Teaching
- Design assistance (for optimization of details, **not** for full design)

The model is not the final answer to biological phosphorus removal models. It is a compromise between complexity and simplicity, and between the many viewpoints on what the correct model should look like. It should be used as a conceptual platform for further model development.

ASM2 is an extension of the **Activated Sludge Model No. 1 (ASM1)**, and uses the concepts incorporated in that model (Henze *et al.*, 1987). **ASM1** has proved to be an excellent tool for modelling nitrification-denitrification processes and has initiated further research in modelling and wastewater characterization. It is hoped that **ASM2** will serve a similar function.

1.1 Background

The strong movement towards effluent criteria for both nitrogen and phosphorus has created a need for a tool to model biological phosphorus removal processes. In the planning and design of new nutrient removal treatment plants a dynamic model is useful because it allows a multitude of scenarios to be tested (low temperature, process start-up, process disturbances, crisis management).

For modern nutrient removal plants, optimization is complicated, because of the many interacting processes. A model is a valuable tool for optimizing the operation. For nutrient removal treatment plants with operational problems, a model is useful for evaluating and implementing new operational procedures.

1.2 Conceptual approach

An attempt has been made to limit the number of processes used in the model. The aim has, however, been to produce a model that can reasonably describe the many different process configurations which are used for biological phosphorus removal. This has resulted in the present level of complexity. In specific cases, it will be possible to reduce the complexity of the model by omitting processes that do not play a significant role, without interfering with the predictive power of the model.

The kinetics and stoichiometry used to describe the processes have been chosen as simply as possible, mainly based on Monod kinetics for all components that can influence the reaction rates. Monod kinetics allow for smooth transitions of the processes, as experience has shown. Kinetics and stoichiometry are presented using the matrix notation, which is the only possible method to overview the complex transformations among the components. The matrix notation also allows control of the continuity of the stoichiometric coefficients and thus ensures that mass balances in the calculations are correctly maintained.

1.3 Limitations

ASM2 has many limitations. It is based on information from municipal wastewater treatment processes. Incorporated in a computer program, every conceivable case can be simulated. However, it must be remembered that results from cases beyond the normal range of experience might be useful in the development of new processes, but that the calculated results may not always be valid.

1.4 Symbols and definitions

Definitions of all components can be found in Chapter 2. Symbols and default values are given in Tables 3.1, 3.2 and 3.3.

1.5 Final note

To present a model, such as the one in this report, will no doubt stimulate debate, as was the case for **ASM1**. Of importance is that it will focus research on the weakest parts of the model, and thus initiate a process that will develop nutrient removal systems, their operational strategies, experimental techniques and mathematical modelling further.

Models, however complex they might be, will still help to organize the thinking of researchers and thus reduce experimental waste.

2. The Activated Sludge Model No. 2

The **Activated Sludge Model No. 2 (ASM2)** is an extension of the **Activated Sludge Model No. 1 (ASM1). ASM2** is more complex and includes many more components which are required in order to characterize the wastewater as well as the activated sludge. Additional biological processes are included, primarily in order to deal with biological phosphorus removal. The most significant change from **ASM1** to **ASM2** is the fact that the biomass now has cell internal structure, and therefore its concentration cannot simply be described with the distributed parameter X_{BM}. This is a prerequisite in order to include biological phosphorus removal in the model.

In addition to the biological processes, **ASM2** includes two 'chemical processes', which may be used to model chemical precipitation of phosphorus.

Whereas **ASM1** was based entirely on COD for all particulate organic material, as well as the total concentration of the activated sludge, **ASM2** includes poly-phosphates, a fraction of the activated sludge which is of prime importance for the performance of the activated sludge system, but which does not exert any COD. For this reason, the possibility of including total suspended solids (TSS) in the model is introduced. TSS also allow for inclusion of mineral particulate solids in the influent to treatment plants, as well as generation of such solids in the context of precipitation of phosphorus.

ASM2 is introduced here in a form which is more complex than a basic version, which could still predict many of the phenomena within a biological nutrient removal plant. The complex model as presented may easily be simplified by eliminating those components which do not have a dominant effect upon the kinetics of the processes, or the aspects of performance of the plant which are of interest.

ASM2 does not distinguish between the composition (cell internal structure) of individual cells but considers only the average composition of the biomass. Since each cell has a different history, its composition will typically deviate from the population average (e.g. it may not contain storage products whereas the the average cell still has storage products available). This is of importance because kinetic expressions used in **ASM2** are non-linear, and therefore average behaviour may not necessarily be predicted from average properties. In view of the additional problems that population models would introduce, the Task Group took the pragmatic decision to accept these problems and to propose **ASM2** based on average properties of the population.

2.1 Components in the model

All symbols for model components distinguish between soluble '$S_?$' and particulate '$X_?$'. Within the activated sludge systems, particulate components, $X_?$, are assumed to be associated with the activated sludge (flocculated onto the activated sludge). They can be concentrated by sedimentation/thickening in clarifiers whereas soluble components, $S_?$, will only be transported with the water.

All particulate model components, $X_?$, must be electrically neutral (no ionic charges), soluble components, $S_?$ may carry ionic charges.

Soluble and particulate components may not necessarily be differentiated by filtration through 0.45 μm membrane filters as is frequently assumed in the technical literature. Some of these components are defined by their interaction with the biomass and require bioassays for their analysis (see Chapter 4 for details).

All components are assumed to be homogeneous and distributed throughout the systems of interest.

2.1.1 Definition of soluble components, '$S_?$'

S_A [M(COD) L^{-3}]: Fermentation products, considered to be acetate. Since fermentation is included in the biological processes, the fermentation products must be modelled separately from other soluble organic materials. They are end-products of fermentation. For all stoichiometric computations, it is assumed that S_A is equal to acetate, in reality a whole range of other fermentation products is possible.

S_{ALK} [mol(HCO_3^-) L^{-3}]: Alkalinity of the wastewater. Alkalinity is used to approximate the continuity of electrical charges in biological reactions. Alkalinity is introduced in order to obtain an early indication of possible low pH conditions, which might inhibit some biological processes. For all stoichiometric computations, S_{ALK} is assumed to be bicarbonate, HCO_3^- only.

S_F [M(COD) L^{-3}]: Fermentable, readily bio-

degradable organic substrates. This fraction of the soluble COD is directly available for biodegradation by heterotrophic organisms. It is assumed that S_F may serve as a substrate for fermentation, therefore it does not include fermentation products.

S_I [M(COD) L^{-3}]: **Inert soluble organic material.** The prime characteristic of S_I is that these organics cannot be further degraded in the treatment plants dealt with in this report. This material is assumed to be part of the influent and it is also assumed to be produced in the context of hydrolysis of particulate substrates X_S.

S_{N_2} [M(N) L^{-3}]: **Dinitrogen, N_2.** S_{N_2} is assumed to be the only product of denitrification. S_{N_2} may be subject to gas exchange, parallel with oxygen, S_{O_2}.

S_{NH_4} [M(N) L^{-3}]: **Ammonium plus ammonia nitrogen.** For the balance of the electrical charges, S_{NH_4} is assumed to be all NH_4^+.

S_{NO_3} [M(N) L^{-3}]: **Nitrate plus nitrite nitrogen (NO_3^- + NO_2^--N).** S_{NO_3} is assumed to include nitrate as well as nitrite nitrogen, since nitrite is not included as a separate model component. For all stoichiometric computations (COD conservation), S_{NO_3} is considered to be NO_3^--N only.

S_{O_2} [M(O_2) L^{-3}]: **Dissolved oxygen.** Dissolved oxygen may be subject to gas exchange.

S_{PO_4} [M(P) L^{-3}]: **Inorganic soluble phosphorus, primarily ortho-phosphates.** For the balance of electrical charges, it is assumed that S_{PO_4} consists of 50% $H_2PO_4^-$ and 50% HPO_4^{2-}, independent of pH.

S_S [M(COD) L^{-3}]: **Readily biodegradable substrate.** This component was introduced in **ASM1**. In **ASM2**, it is replaced by the sum of S_F + S_A.

2.1.2 Definition of particulate components '$X_?$'

X_{AUT} [M(COD) L^{-3}]: **Nitrifying organisms.** Nitrifying organisms are responsible for nitrification; they are obligate aerobic, chemo-litho-autotrophic. It is assumed that nitrifiers oxidize ammonium S_{NH_4} directly to nitrate S_{NO_3} (nitrifiers include both *Nitrosomonas* and *Nitrobacter*).

X_H [M(COD) L^{-3}]: **Heterotrophic organisms.** These organisms are assumed to be the 'all-rounder' heterotrophic organisms, they may grow aerobically and anoxically (denitrification) and be active anaerobically (fermentation). They are responsible for hydrolysis of particulate substrates X_S and can use all degradable organic substrates under all relevant environmental conditions.

X_I [M(COD) L^{-3}]: **Inert particulate organic material.** This material is not degraded within the systems of interest. It is flocculated onto the activated sludge. X_I may be a fraction of the influent or may be produced in the context of biomass decay.

X_{MeOH} [M(TSS) L^{-3}]: **Metal-hydroxides.** This component stands for the phosphorus-binding capacity of possible metal-hydroxides, which may be in the wastewater or may be added to the system. For all stoichiometric computations, it is assumed that this component is composed of $Fe(OH)_3$. It is possible to 'replace' this component with other reactants; this would require adaptation of the stoichiometric and kinetic information.

X_{MeP} [M(TSS) L^{-3}]: **Metal-phosphate, $MePO_4$.** This component results from binding phosphorus to the metal-hydroxides. For all stoichiometric computations, it is assumed that this component is composed of $FePO_4$. It is possible to 'replace' this component with other precipitation products; this would require adaptation of the stoichiometric and kinetic information.

X_{PAO} [M(COD) L^{-3}]: **Phosphate-accumulating organisms: PAO.** These organisms are assumed to be representative for all types of poly-phosphate-accumulating organism. The concentration of X_{PAO} does not include the cell internal storage products X_{PP} and X_{PHA}, but only the 'true' biomass.

X_{PHA} [M(COD) L^{-3}]: **A cell internal storage product of phosphorus-accumulating organisms, PAO.** It includes poly-hydroxy-alkanoates (PHA), glycogen, etc. It occurs only associated with X_{PAO}; it is, however, not included in the mass of X_{PAO}. X_{PHA} cannot be directly compared with analytically measured PHA or glycogen concentrations; X_{PHA} is only a functional component required for modelling but not directly identifiable chemically. X_{PHA} may, however, be recovered in COD analysis, where it must satisfy COD continuity. For stoichiometric considerations, PHA is assumed to have the chemical composition of poly-hydroxy-butyrate $(C_4H_6O_2)_n$.

X_{PP} [M(P) L^{-3}]: **Poly-phosphate.** Poly-phosphate is a cell internal inorganic storage product of PAO. It occurs only associated with X_{PAO}; it is, however, not included in the mass of X_{PAO}. It is part of the particulate phosphorus and may be analytically observed. For stoichiometric considerations, poly-phosphates are assumed to have the composition of $(K_{0.33}Mg_{0.33}PO_3)_n$.

X_S [M(COD) L^{-3}]: **Slowly biodegradable substrates.** Slowly biodegradable substrates are high molecular weight, colloidal and particulate organic substrates which must undergo cell external hydrolysis before they are available for degradation. It is assumed that the products of hydrolysis (S_F) may be fermented.

X_{TSS} [M(TSS) L^{-3}]: **Total suspended solids, TSS.** Total suspended solids are introduced into the biokinetic models in order to compute their concentration via stoichiometry. Since phosphorus removal and precipitation introduce mineral fractions into the activated sludge, prediction of TSS becomes important.

2.2 Basis for the introduction of ASM2

2.2.1 Matrix notation

The Task Group introduced matrix notation for the presentation of biokinetic models in its report on the **ASM1**. The same concept will be used for the introduction of **ASM2**. It is assumed that the reader is familiar with this way of presenting biokinetics.

As a short summary: the components which are considered in the model and the transformation processes are characterized with the indices i and j respectively. Stoichiometric coefficients are presented in the form of a stoichiometric matrix ν_{ji}. The process rate equations form a vector ρ_j. The rate of production of the component i, r_i [M_i L^{-3} T^{-1}], in all parallel processes may then be computed from the sum:

$$r_i = \sum \nu_{ji} \cdot \rho_j \quad \text{over all processes } j \tag{2.1}$$

Within the stoichiometric matrix one stoichiometric coefficient, ν_{jk}, of each process j may be chosen as dimensionless with the value of +1 or -1. For all other stoichiometric coefficients algebraic equations may be given, which introduce continuity principles into the determination of stoichiometric coefficients. Alternatively ν_{ji} may be given in the form of absolute values with the dimension M_i M_k^{-1}, where M_k is the unit mass of the component k upon which stoichiometry is based (the component which has ν_{jk} = +1 or -1).

2.2.2 Continuity equations

Continuity equations are the mathematical equivalent of the principle that in chemical reactions, elements, electrons (or COD) and net electrical charges may neither be formed nor destroyed.

The stoichiometry of **ASM1** is implicitly based on three continuity considerations for COD, electrical charges and nitrogen. **ASM2** adds phosphorus continuity to these three. Further, an equation is introduced which converts the different solid components $X_?$ from their unit of measurement, to total suspended solids, X_{TSS}.

A continuity equation, which is valid for all processes j and all materials c subject to continuity, may be written as:

$$\sum \nu_{ji} \cdot i_{ci} = 0 \quad \text{over all components } i \tag{2.2}$$

where

ν_{ji} = stoichiometric coefficient for component i in process j [M_i M_k^{-1}],

i_{ci} = conversion factor to convert the units of component i to the units of the material c, to which continuity is to be applied [M_c M_i^{-1}].

Each continuity equation contains *a priori* information and may be applied to each process. Each continuity equation allows the prediction of one stoichiometric coefficient without performing an experiment, provided the other coefficients are known.

In **ASM2**, these equations are used to estimate

Table 2.1. Conversion factors i_{ci} to be applied in the continuity equations of ASM2. Missing values are equal to 0. The units of i_{ci} are M_c M_i^{-1}, e.g. $i_{N,2} = i_{NSF}$ g N (g COD)$^{-1}$ or $i_{Charge,3}$ = -1/64 moles$^+$ g COD^{-1}.

Index c: Continuity for			COD	N	P	Charge	Mass
Factor:			$i_{COD,i}$	$i_{N,i}$	$i_{P,i}$	$i_{Charge,i}$	$i_{TSS,i}$
i	Component	Units	g COD	g N	g P	mole$^+$	g TSS
1	S_{O_2}	g O_2	-1				
2	S_F	g COD	1	i_{NSF}	i_{PSF}		
3	S_A	g COD	1			-1/64	
4	S_{NH_4}	g N		1		+1/14	
5	S_{NO_3}	g N	-64/14	1		-1/14	
6	S_{PO_4}	g P			1	-1.5/31	
7	S_I	g COD	1	i_{NS_I}	i_{PS_I}		
8	S_{ALK}	mole HCO_3^-				-1	
9	S_{N_2}	g N	-24/14	1			
10	X_I	g COD	1	i_{NX_I}	i_{PX_I}		i_{TSSX_I}
11	X_S	g COD	1	i_{NX_S}	i_{PX_S}		i_{TSSX_S}
12	X_H	g COD	1	i_{NBM}	i_{PBM}		i_{TSSBM}
13	X_{PAO}	g COD	1	i_{NBM}	i_{PBM}		i_{TSSBM}
14	X_{PP}	g P			1		3.23
15	X_{PHA}	g COD	1				0.60
16	X_{AUT}	g COD	1	i_{NBM}	i_{PBM}		i_{TSSBM}
17	X_{TSS}	g TSS					-1[a)]
18	X_{MeOH}	g TSS					1
19	X_{MeP}	g TSS			0.205		1

a) Since TSS are counted twice, this factor must be negative.

All absolute numbers are obtained based on the chemical composition of the component (see definition of component). All factors i_{ci} are model parameters and must be estimated from experiments.

Table 2.2. Stoichiometry of hydrolysis processes. The stoichiometric parameters are defined in Table 3.2.

	Process	S_F	S_{NH4}	S_{PO4}	S_I	S_{ALK}	X_S	X_{TSS}
1	Aerobic hydrolysis	$1\text{-}f_{SI}$	v_{1,NH_4}	$v_{1,PO4}$	f_{SI}	$v_{1,ALK}$	-1	$v_{1,TSS}$
2	Anoxic hydrolysis	$1\text{-}f_{SI}$	v_{2,NH_4}	$v_{2,PO4}$	f_{SI}	$v_{2,ALK}$	-1	$v_{2,TSS}$
3	Anaerobic hydrolysis	$1\text{-}f_{SI}$	v_{3,NH_4}	$v_{3,PO4}$	f_{SI}	$v_{3,ALK}$	-1	$v_{3,TSS}$

The stoichiometric coefficients for S_{NH_4}, S_{PO_4}, S_{ALK} and X_{TSS} may be computed from the Continuity Equation 2.2 with the aid of Table 2.1. As an example $v_{1,PO_4} = -\,[(1\text{-}f_{SI}) \cdot i_{PS_F} + f_{SI} \cdot i_{PS_I} - 1 \cdot i_{PX_S}] / 1$.

the stoichiometric coefficients of S_{O_2} (S_{NO_3} and S_{N_2} in denitrification) from COD, S_{NH_4} from nitrogen, S_{PO_4} from phosphorus, S_{ALK} from charge and X_{TSS} from total solids continuity. Table 2.1 is a summary of the conversion factors i_{ci} which must be applied in Equation 2.2. These conversion factors are, wherever possible, obtained from chemical stoichiometry. 'COD' as a conservative property is defined as closely as possible to the analytically obtained COD. Examples are:

$$i_{COD,5} = -\,64\ g\ O_2 / 14\ g\ NO_3^-\text{-}N \text{ from:}$$

$$NO_3^- + H_2O + 2\,H^+ \rightarrow NH_4^+ + 2\,O_2$$

Or, one mole of nitrate (14 g N) has a negative oxygen demand ('liberates oxygen') of two moles of oxygen (64 g O_2). Similar arguments lead to:

$$i_{COD,9} = -\,24\ g\ O_2 / 14\ g\ N_2 \text{ from:}$$

$$2\,N_2 + 6\,H_2O + 4\,H^+ \rightarrow 4\,NH_4^+ + 3\,O_2$$

All conversion factors given with absolute numbers in Table 2.1 may be obtained from chemical stoichiometry, based on the definition of the compounds. All factors identified with a symbol i_{ci} must be obtained from chemical analysis.

As an example, the stoichiometric coefficient for component 2 (i = 2) in the third process (j = 3) may be obtained from the continuity equation for COD based on Equation 2.2 according to:

$$v_{3,2} = -\,(v_{3,1} \cdot i_{COD,1} + v_{3,3} \cdot i_{COD,3} + \ldots + v_{3,n} \cdot i_{COD,n}) / i_{COD,2}$$

or

$$v_{3,2} = -\,[\,\sum_i (v_{3,i} \cdot i_{COD,i}) - v_{3,2} \cdot i_{COD,2}] / i_{COD,2}$$

The introduction of the continuity equations in an abstract form may at first appear to be complicated. However, the concept is directed towards its application in computer programs and helps to simplify the development of program code.

2.3 Biological processes, stoichiometry and kinetics

The biological processes of **ASM2** are introduced here. A full stoichiometric matrix using typical stoichiometric coefficients is presented in Chapter 3 (Table 3.4).

2.3.1 Biological processes, general remarks

Microorganisms have a complex cell internal structure and respond to different environmental conditions with adjustment of this structure. A frequently observed phenomenon is unbalanced growth, a situation where not all fractions of the cells are reproduced at an equal rate. Modelling such shifts of cell internal structure would require modelling of the different fractions of the biomass, a task which would be most fruitful if the behaviour of axenic cultures were described. Here, only three groups of microorganisms represent a vast variety of unknown species; each biological process described in **ASM2** represents a large number of processes which act upon a variety of substances, which in the model are summarized in terms of COD.

Process descriptions in **ASM2** are therefore based on the average behaviour of these different microorganisms, and are described as balanced growth processes would be modelled.

2.3.2 Hydrolysis processes

Many high molecular weight, colloidal or particulate organic substrates cannot be utilized directly by microorganisms. These substrates must be made available by cell external enzymatic reactions which are called hydrolysis processes. It is unclear whether the products of hydrolysis ever exist in true solution or whether they are taken up directly by the organisms which catalyse hydrolysis. Typically hydrolysis processes are considered to be surface reactions, which occur in close contact between the organisms which provide the hydrolytic enzymes and the slowly biodegradable substrates themselves.

There is experimental evidence that hydrolysis reactions depend on the available electron acceptors, therefore three hydrolysis processes are distinguished in **ASM2**. It is, however, a difficult task to estimate hydrolysis rate constants under different electron acceptor conditions.

1. Aerobic hydrolysis of slowly biodegradable substrate characterizes hydrolysis under aerobic conditions ($S_{O_2} > 0$).
2. Anoxic hydrolysis of slowly biodegradable substrate characterizes hydrolysis under anoxic conditions ($S_{O_2} \approx 0$, $S_{NO_3} > 0$). This process is typically slower than aerobic hydrolysis.
3. Anaerobic hydrolysis of slowly biodegradable substrate characterizes hydrolysis under anaerobic conditions ($S_{O_2} \approx 0$, $S_{NO_3} \approx 0$). This process is not well characterized and is probably slower than aerobic hydrolysis. Its rate remains to be studied.

Table 2.2 summarizes the stoichiometry of the hydrolysis processes. It is assumed that slowly biodegradable substrate X_S is degraded to readily

Table 2.3. Stoichiometry of the growth and decay processes of facultative heterotrophic organisms X_H. The stoichiometric parameters are defined in Table 3.2. Stoichiometry for S_{O_2}, S_{NH_4}, S_{PO_4}, S_{ALK} and X_{TSS} may be computed from continuity.

	Process	S_{O_2}	S_F	S_A	S_{NO_3}	S_{N_2}	X_I	X_S	X_H
4	Aerobic growth on S_F	$1-\frac{1}{Y_H}$	$\frac{-1}{Y_H}$						1
5	Aerobic growth on S_A	$1-\frac{1}{Y_H}$		$\frac{-1}{Y_H}$					1
6	Anoxic growth on S_F Denitrification		$\frac{-1}{Y_H}$		$-\frac{1-Y_H}{2.86\,Y_H}$	$\frac{1-Y_H}{2.86\,Y_H}$			1
7	Anoxic growth on S_A Denitrification			$\frac{-1}{Y_H}$	$-\frac{1-Y_H}{2.86\,Y_H}$	$\frac{1-Y_H}{2.86\,Y_H}$			1
8	Fermentation		-1	1					
9	Lysis						f_{X_I}	$1-f_{X_I}$	-1

degradable substrate S_F whereby a small fraction f_{S_I} of inert organic material S_I is released. The stoichiometric coefficients for S_{NH_4}, S_{PO_4} and S_{ALK} may be computed from Continuity Equation 2.2. These three coefficients are typically positive.

The proposed rate equations for the hydrolysis processes 1–3 are presented in Table 2.7. They are similar to those of **ASM1**: hyperbolic switching functions for S_{O_2} and S_{NO_3} consider the environmental conditions; a surface-limited reaction (X_S/X_H) / (K_X+X_S/X_H) is assumed for the hydrolysis process itself. It is proposed that only heterotrophic organisms may catalyse hydrolysis. Typically hydrolysis is slower under denitrifying or anaerobic (fermentation) than under aerobic conditions. The rate for anoxic and anaerobic hydrolysis is therefore reduced by the factors η_{NO_3} and η_{fe} respectively.

The hydrolysis of particulate, biodegradable organic nitrogen is included as a separate process in **ASM1** but not in **ASM2**. This process is necessary if the nitrogen content of X_S is variable. In order to simplify **ASM2**, it is assumed that X_S contains a constant fraction of nitrogen i_{NX_S} and phosphorus i_{PX_S}. Without this simplifying assumption, six more hydrolysis processes and two more particulate components would be required.

The process of ammonification is included in **ASM1** in order to describe the release of ammonium, S_{NH_4}, from soluble, biodegradable organic nitrogen. In **ASM2** it is assumed that the fermentable substrates, S_F, contain a constant fraction of nitrogen and phosphorus, i_{NS_F} and i_{PS_F} respectively. This allows the process of ammonification to be ignored. Without this simplifying assumption, two more processes (ammonification as well as phosphatification, the release of phosphate S_{PO_4} from an organic fraction), and two more components (soluble, degradable organic nitrogen and phosphorus) would have to be introduced.

2.3.3 Processes of facultative heterotrophic organisms

The heterotrophic organisms X_H are responsible for the hydrolysis of slowly biodegradable substrate X_S (see above), the aerobic degradation of fermentable organic substrates S_F and of fermentation products S_A (aerobic growth), anoxic oxidation of S_F and S_A and reduction of nitrate S_{NO_3} (denitrification), and anaerobic fermentation of S_F to S_A. In addition these organisms are subject to decay and lysis. The stoichiometry and the kinetics of the processes described below are presented in Tables 2.3 and 2.7 respectively.

4. and 5. Aerobic growth of heterotrophic organisms on fermentable substrates S_F and on fermentation products S_A. These processes are modelled as two parallel processes, which consume the two degradable organic substrates S_F and S_A. For both processes identical growth rates μ_m and yield coefficients Y_H are assumed. The rate equations are designed such that the maximum specific growth rate of the heterotrophic organisms does not increase above μ_m even if both substrates, S_F and S_A, are present in high concentrations. These processes require oxygen, S_{O_2}, nutrients, S_{NH_4} and S_{PO_4}, and possibly alkalinity, S_{ALK}, and they produce suspended solids, X_{TSS}.

6. and 7. Anoxic growth of heterotrophic organisms on fermentable substrates, S_F, and on fermentation products, S_A; denitrification. These two processes are similar to the aerobic growth processes, but they require nitrate, S_{NO_3}, as the electron acceptor rather than oxygen. The stoichiometry for nitrate is computed based on the assumption that all nitrate, S_{NO_3}, is reduced to dinitrogen, S_{N_2}. Denitrification releases alkalinity, the stoichiometry of which is predicted from continuity. Denitrification is assumed to be inhibited by oxygen S_{O_2} and the

Table 2.4. Stoichiometry of the processes of phosphorus-accumulating organisms, PAO. The stoichiometric parameters are defined in Table 3.2. Stoichiometry for S_{O_2}, S_{NH_4}, S_{PO_4}, S_{ALK} and X_{TSS} may be computed from continuity.

	Process	S_{O_2}	S_A	S_{PO4}	X_I	X_S	X_{PAO}	X_{PP}	X_{PHA}
10	Storage of X_{PHA}		-1	Y_{PO_4}				$-Y_{PO_4}$	1
11	Storage of X_{PP}	$-Y_{PHA}$		-1				1	$-Y_{PHA}$
12	Aerobic growth of X_{PAO}	$1-\frac{1}{Y_H}$		$-i_{PBM}$			1		$-\frac{1}{Y_H}$
13	Lysis of X_{PAO}			v_{13,PO_4}	f_{X_I}	$1-f_{X_I}$	-1		
14	Lysis of X_{PP}			1				-1	
15	Lysis of X_{PHA}		1						-1

maximum growth rate μ_m is reduced relative to its value under aerobic conditions, by the factor η_{NO_3}. This accounts for the fact that not all heterotrophic organisms X_H may be capable of denitrification or that denitrification may only proceed at a reduced rate.

8. Fermentation. Under anaerobic conditions ($S_{O_2} \approx 0$, $S_{NO_3} \approx 0$) it is assumed that heterotrophic organisms are capable of fermentation, whereby readily biodegradable substrates S_F are transformed into fermentation products S_A. Although this process may possibly cause growth of heterotrophic organisms, it is introduced here as a simple transformation process. A growth process would require more complex kinetics, more kinetic and stoichiometric parameters which are difficult to obtain, and possibly different yield coefficients for S_F and S_A in processes 4 to 7. Fermentation releases negatively charged fermentation products, S_A, and therefore has a requirement for alkalinity, S_{ALK}. This is predicted from continuity.

Fermentation is a process which, up to now, has not been well characterized. Little is known about the kinetics of this process, which may lead to a large range of kinetic parameters for modelling experimental results. Reliable application of **ASM2** requires that research is directed towards characterizing what is described here with the process of fermentation.

9. Lysis of heterotrophic organisms. This process represents the sum of all decay processes of the heterotrophic organisms. It is modelled in analogy to **ASM1**; its rate is independent of environmental conditions.

2.3.4 Processes of phosphorus-accumulating organisms

Some organisms, X_{PAO}, are known for their potential to accumulate phosphorus in the form of poly-phosphate X_{PP}. Currently these organisms are not well characterized; historically it was assumed that they would all be part of the *Acinetobacter* genus. However, today it is clear that *Acinetobacter* contribute to, but do not always dominate, biological phosphorus removal. Initially it was assumed that phosphorus-accumulating organisms, PAO, could not denitrify; now evidence has become available that some of them can denitrify. Phosphate release is sometimes slower in the presence of nitrate; this observation is not predicted with **ASM2**. Glycogen is found to be an important carbon storage material of PAO but is not considered in **ASM2** in order to reduce model complexity. All these restrictions lead to limitations of the applicability of the model which will be discussed later.

The greater the attempts to characterize PAO, the more complex this group of organisms becomes. The Task Group is well aware that the time has come when biological phosphorus removal is being designed and used in actual plants. The introduction of a very detailed mechanistic model for the processes responsible for biological phosphorus removal is, however, premature. The Task Group therefore has chosen to suggest a simple model, which allows prediction of biological phosphorus removal, but does not yet include all observed phenomena. The model proposed may be the base for further development.

The following model for the behaviour of phosphorus-accumulating organisms, X_{PAO}, assumes that these organisms cannot denitrify and that they can only grow on cell internal stored organic materials, X_{PHA}. Both these assumptions are very severe restrictions of **ASM2** and may lead to further extensions. The stoichiometry and the kinetics of the processes described below are presented in Tables 2.4 and 2.7 respectively.

10. Storage of X_{PHA}. It is assumed that PAO may release phosphate, S_{PO_4} from poly-phosphate, X_{PP}, and utilize the energy which becomes available from the hydrolysis of X_{PP}, in order to store cell external fermentation products S_A in the form of cell internal organic storage material X_{PHA}. The process is primarily observed under anaerobic conditions. However, since the process has also been reported to occur under aerobic and anoxic conditions, the kinetic expression does not include inhibition terms for S_{O_2} and S_{NO_3}. Experimental observation of this process is easy if the release of phosphorus is

Table 2.5. Stoichiometry of the growth and decay processes of nitrifying organisms. The stoichiometric parameters are defined in Table 3.2. Stoichiometry for S_{O_2}, S_{NH_4}, S_{PO_4}, S_{ALK} and X_{TSS} may be computed from continuity.

	Processes	S_{O2}	S_{NH4}	S_{NO3}	S_{PO4}	X_I	X_S	X_{AUT}
16	Aerobic growth of X_{AUT}	$-\frac{4.57-Y_A}{Y_A}$	$-i_{NBM}-\frac{1}{Y_A}$	$\frac{1}{Y_A}$	$-i_{PBM}$			1
17	Lysis		$v_{17,NH4}$		$v_{17,PO4}$	f_{X_I}	$1-f_{X_I}$	-1

observed rather than the organics which are stored. Experience indicates, however, that the rate of storage of organics is relatively constant, whereas the release of phosphorus varies, indicating a variable stoichiometric relationship. The base for the stoichiometry of this process was therefore chosen to be the organics which are taken up, S_A and X_{PHA}. Reliable estimation of the rate constant, q_{PHA}, and the stoichiometric parameter, Y_{PO4}, requires independent measurement of both S_A removal and S_P release.

11. Storage of poly-phosphate. Storage of ortho-phosphate, S_{PO4}, in the form of cell internal poly-phosphates, X_{PP}, requires the PAO to obtain energy, which may be gained from the respiration of X_{PHA}. The regeneration of poly-phosphates is a requirement for the growth of PAO, because the organic substrates, S_A, are stored only upon the release of poly-phosphate. Storage of X_{PP} is observed to stop if the phosphorus content of the PAO becomes too high. This observation leads to an inhibition term of X_{PP} storage, which becomes active as the ratio X_{PP}/X_{PAO} approaches the maximum allowable value of K_{MAX}.

12. Growth of phosphorus-accumulating organisms. These organisms are assumed to grow only at the expense of cell internal organic storage products X_{PHA}. As phosphorus is continuously released by the lysis of X_{PP}, it is possible to assume that the organisms consume ortho-phosphate, S_{PO4}, as a nutrient for the production of biomass. Growth of PAO is modelled as an obligate aerobic process. It is known that PAO may grow at the expense of soluble substrates (e.g. S_A), but it is unlikely that such substrates ever become available under aerobic conditions in a biological nutrient removal plant. The Task Group therefore suggests this possibility is ignored at this time.

13. 14. and 15. Lysis of phosphorus-accumulating organisms and their storage products. Death, endogenous respiration and maintenance all result in a loss or decay of all fractions of PAO. Since the storage products X_{PP} and X_{PHA} are accounted for separately from the biomass X_{PAO}, all three components must be subject to separate decay processes. **ASM2** includes three lysis processes which are all first-order relative to the component which is lost. If all three rate constants are equal, the composition of the organisms does not change due to decay. There is experimental evidence that X_{PP} decays faster than X_{PAO} and X_{PHA}. This additional loss of poly-phosphates may be predicted by the choice of an increased rate, b_{PP}, for the lysis of this component. The products of lysis are chosen in analogy to the lysis of heterotrophic organisms; storage products are assumed to decay to ortho-phosphate S_{PO4} and fermentation products S_A.

2.3.5 Nitrification processes

Nitrification is assumed to be a one-step process, from ammonium S_{NH4} directly to nitrate S_{NO3}. The intermediate component, nitrite, is not included as a model component. In the context of nitrification, modelling nitrite production and consumption would be relatively easy. However, nitrite is also produced and consumed in the context of denitrification where the Task Group felt that the required addition to the model complexity does not warrant its inclusion at the present

Table 2.6. Stoichiometry and kinetics of the processes describing simultaneous precipitation of phosphorus. The absolute values for stoichiometry and kinetics are based on the assumption that $FeCl_3$ is used to precipitate S_{PO_4} in the form of $FePO_4$ + $Fe(OH)_3$. Stoichiometry for S_{ALK} and X_{TSS} may be computed from continuity.

	Processes	S_{PO4}	S_{ALK}	X_{MeOH}	X_{MeP}	X_{TSS}
	Stoichiometry:					
18	Precipitation	-1	$v_{19,ALK}$	-3.45	4.87	1.42
19	Redissolution	1	$v_{20,ALK}$	3.45	-4.87	-1.42
	Kinetics:	*Process rate* ρ_j:				
18	Precipitation	$k_{PRE} \cdot S_{PO4} \cdot X_{MeOH}$		k_{PRE} = 1 m^3 (g Fe $(OH)_3)^{-1}$ d^{-1}		
19	Redissolution	$k_{RED} \cdot X_{MeP}$		k_{RED} = 0.6 d^{-1}		

Table 2.7. Process rate equations for ASM2. The kinetic parameters are defined in Table 3.3.

Process	Rate equation ρ
Hydrolysis processes:	
1 Aerobic hydrolysis	$K_h \cdot \frac{S_{O_2}}{K_{O_2}+S_{O_2}} \cdot \frac{X_S/X_H}{K_X+X_S/X_H} \cdot X_H$
2 Anoxic hydrolysis	$K_h \cdot \eta_{NO_3} \cdot \frac{K_{O_2}}{K_{O_2}+S_{O_2}} \cdot \frac{S_{NO_3}}{K_{NO_3}+S_{NO_3}} \cdot \frac{X_S/X_H}{K_X+X_S/X_H} \cdot X_H$
3 Anaerobic hydrolysis	$K_h \cdot \eta_{fe} \cdot \frac{K_{O_2}}{K_{O_2}+S_{O_2}} \cdot \frac{K_{NO_3}}{K_{NO_3}+S_{NO_3}} \cdot \frac{X_S/X_H}{K_X+X_S/X_H} \cdot X_H$
Heterotrophic organisms: X_H	
4 Growth on fermentable substrates, S_F	$\mu_H \cdot \frac{S_{O_2}}{K_{O_2}+S_{O_2}} \cdot \frac{S_F}{K_F+S_F} \cdot \frac{S_F}{S_F+S_A} \cdot \frac{S_{NH_4}}{K_{NH_4}+S_{NH_4}} \cdot \frac{S_{PO_4}}{K_P+S_{PO_4}} \cdot \frac{S_{ALK}}{K_{ALK}+S_{ALK}} \cdot X_H$
5 Growth on fermentation products, S_A	$\mu_H \cdot \frac{S_{O_2}}{K_{O_2}+S_{O_2}} \cdot \frac{S_A}{K_A+S_A} \cdot \frac{S_A}{S_F+S_A} \cdot \frac{S_{NH_4}}{K_{NH_4}+S_{NH_4}} \cdot \frac{S_{PO_4}}{K_P+S_{PO_4}} \cdot \frac{S_{ALK}}{K_{ALK}+S_{ALK}} \cdot X_H$
6 Denitrification on fermentable substrates, S_F	$\mu_H \cdot \eta_{NO_3} \cdot \frac{K_{O_2}}{K_{O_2}+S_{O_2}} \cdot \frac{S_F}{K_F+S_F} \cdot \frac{S_F}{S_F+S_A} \cdot \frac{S_{NH_4}}{K_{NH_4}+S_{NH_4}} \cdot \frac{S_{NO_3}}{K_{NO_3}+S_{NO_3}} \cdot \frac{S_{ALK}}{K_{ALK}+S_{ALK}} \cdot \frac{S_{PO_4}}{K_P+S_{PO_4}} \cdot X_H$
7 Denitrification on fermentation products, S_A	$\mu_H \cdot \eta_{NO_3} \cdot \frac{K_{O_2}}{K_{O_2}+S_{O_2}} \cdot \frac{S_A}{K_A+S_A} \cdot \frac{S_A}{S_F+S_A} \cdot \frac{S_{NH_4}}{K_{NH_4}+S_{NH_4}} \cdot \frac{S_{NO_3}}{K_{NO_3}+S_{NO_3}} \cdot \frac{S_{ALK}}{K_{ALK}+S_{ALK}} \cdot \frac{S_{PO_4}}{K_P+S_{PO_4}} \cdot X_H$
8 Fermentation	$q_{fe} \cdot \frac{K_{O_2}}{K_{O_2}+S_{O_2}} \cdot \frac{K_{NO_3}}{K_{NO_3}+S_{NO_3}} \cdot \frac{S_F}{K_{fe}+S_F} \cdot \frac{S_{ALK}}{K_{ALK}+S_{ALK}} \cdot X_H$
9 Lysis	$b_H \cdot X_H$
Phosphorus-accumulating organisms (PAO): X_{PAO}	
10 Storage of X_{PHA}	$q_{PHA} \cdot \frac{S_A}{K_A+S_A} \cdot \frac{S_{ALK}}{K_{ALK}+S_{ALK}} \cdot \frac{X_{PP}/X_{PAO}}{K_{PP}+X_{PP}/X_{PAO}} \cdot X_{PAO}$
11 Storage of X_{PP}	$q_{PP} \cdot \frac{S_{O_2}}{K_{O_2}+S_{O_2}} \cdot \frac{S_{PO_4}}{K_{PS}+S_{PO_4}} \cdot \frac{S_{ALK}}{K_{ALK}+S_{ALK}} \cdot \frac{X_{PHA}/X_{PAO}}{K_{PHA}+X_{PHA}/X_{PAO}} \cdot \frac{K_{MAX}-X_{PP}/X_{PAO}}{K_{IPP}+K_{MAX}-X_{PP}/X_{PAO}} \cdot X_{PAO}$
12 Aerobic growth on X_{PHA}	$\mu_{PAO} \cdot \frac{S_{O_2}}{K_{O_2}+S_{O_2}} \cdot \frac{S_{NH_4}}{K_{NH_4}+S_{NH_4}} \cdot \frac{S_{ALK}}{K_{ALK}+S_{ALK}} \cdot \frac{S_{PO_4}}{K_P+S_{PO_4}} \cdot \frac{X_{PHA}/X_{PAO}}{K_{PHA}+X_{PHA}/X_{PAO}} \cdot X_{PAO}$
13 Lysis of X_{PAO}	$b_{PAO} \cdot X_{PAO} \cdot S_{ALK}/(K_{ALK}+S_{ALK})$
14 Lysis of X_{PP}	$b_{PP} \cdot X_{PP} \cdot S_{ALK}/(K_{ALK}+S_{ALK})$
15 Lysis of X_{PHA}	$b_{PHA} \cdot X_{PHA} \cdot S_{ALK}/(K_{ALK}+S_{ALK})$
Nitrifying organisms (autotrophic organisms): X_{AUT}	
16 Growth	$\mu_{AUT} \cdot \frac{S_{O_2}}{K_{O_2}+S_{O_2}} \cdot \frac{S_{NH_4}}{K_{NH_4}+S_{NH_4}} \cdot \frac{S_{PO_4}}{K_P+S_{PO_4}} \cdot \frac{S_{ALK}}{K_{ALK}+S_{ALK}} \cdot X_{AUT}$
17 Lysis	$b_{AUT} \cdot X_{AUT}$
Simultaneous precipitation of phosphorus with ferric hydroxide $Fe(OH)_3$	
18 Precipitation	$k_{PRE} \cdot S_{PO_4} \cdot X_{MeOH}$
19 Redissolution	$k_{RED} \cdot X_{MeP} \cdot S_{ALK}/(K_{ALK}+S_{ALK})$

time. Modelling nitrite in nitrification but not in denitrification would, however, not be consistent and could lead to erroneous model predictions.

The stoichiometry and the kinetics of the processes described below, are presented in Tables 2.5 and 2.7 respectively.

16. Growth of nitrifying organisms. Nitrifying organisms are obligate aerobic, they consume ammonium as a substrate and a nutrient, and produce nitrate. Nitrification reduces alkalinity. The process is modelled as proposed in **ASM1** with the exception of a phosphorus uptake into the biomass.
17. Lysis of nitrifying organisms. The process of lysis of nitrifiers is modelled in analogy to **ASM1** and to the process of lysis of heterotrophic organisms. Since the decay products of lysis (X_S and ultimately S_F) are available substrates for heterotrophic organisms only, endogenous respiration of nitrifiers becomes manifest as an increased growth and oxygen consumption of heterotrophs. This is in analogy to **ASM1**.

2.4 Chemical precipitation of phosphates

In biological nutrient removal systems, metals, which are naturally present in the wastewater (e.g. Ca^{2+}), together with the high concentration of released soluble ortho-phosphate, S_{PO_4}, may result in chemical precipitation of phosphorus (e.g. in the form of apatite or calcium phosphate).

Further, simultaneous precipitation of phosphorus via the addition of iron or aluminium salts is a very common process for phosphorus removal worldwide. Simultaneous precipitation may be used in combination with biological phosphorus removal if the carbon to phosphorus ratio is unfavourably small.

In order to model the low effluent concentrations of ortho-phosphate, S_{PO_4}, which are observed in practice and which are partly due to chemical precipitation, the Task Group suggests a very simple precipitation model, which may be calibrated for a variety of situations. For this purpose, two processes (precipitation and redissolution) and two more components (X_{MeOH} and X_{MeP}) are added to **ASM2**. If chemical precipitation is not of any interest, these additions may be deleted from the model.

18. and 19. Precipitation and redissolution of phosphate S_{PO_4}. The precipitation model is based on the assumption that precipitation and redissolution are reverse processes, which at steady state would be in equilibrium according to:

$$X_{MeOH} + S_{PO_4} \Leftrightarrow X_{MeP}$$

Precipitation and redissolution may be modelled with the following process rates respectively:

$$\rho_{18} = k_{PRE} \cdot S_{PO_4} \cdot X_{MeOH}$$

$$\rho_{19} = k_{RED} \cdot X_{MeP}$$

If both processes are in equilibrium ($\nu_{18,i} \cdot \rho_{18} = \nu_{19,i} \cdot \rho_{19}$) then an equilibrium constant may be derived as:

$$K_{eq} = \frac{\nu_{19,i} \cdot k_{RED}}{\nu_{18,i} \cdot k_{PRE}} = \frac{S_{PO_4} \cdot X_{MeOH}}{X_{MeP}}$$

Processes 18 and 19 will be introduced here based on the assumption that X_{MeOH} and X_{MeP} are composed of ferric-hydroxide, $Fe(OH)_3$, and ferric-phosphate, $FePO_4$, respectively. This leads to the stoichiometry indicated in Table 2.6. The indicated rates of the processes result in residual ortho-phosphate concentrations, S_P, which at steady state are typical for simultaneous precipitation with the addition of $FeCl_3$. In this case, the addition of Fe^{3+} to the influent of a treatment plant may be modelled by the choice of X_{MeOH} in the influent recognizing that 1 g Fe^{3+} m^{-3} leads to 1.91 g $Fe(OH)_3$ m^{-3} = 1.91 g MeOH m^{-3} (which also increases influent X_{TSS} and decreases influent alkalinity S_{ALK}).

3. Typical wastewater characteristics, kinetic and stoichiometric constants for the Activated Sludge Model No. 2

It is the responsibility of the user of the **Activated Sludge Model No. 2** (**ASM2**) to determine the concentrations of relevant components in the wastewater, as well as the stoichiometric and kinetic parameters which apply to the specific case to be dealt with. Absolute numbers of these parameters are not part of **ASM2**, but are necessary for the application of the model to a specific case.

In this section, the Task Group suggests a list of typical concentrations of model components in a primary effluent as well as a set of model parameters. This neither indicates that **ASM2** is meant to be reliable with these parameters in any case, nor that these parameters are the state of the art. They are merely presented as a reference for testing computer code and a first estimate for the design of possible experiments which are proposed to determine these parameters more accurately.

Table 3.1 contains a list of all model components and typical concentrations in a primary effluent. This wastewater contains a total COD of 260 g COD m^{-3}, a total nitrogen content of 25 g N m^{-3} and approximately 140 g TSS m^{-3}. The analytically measured TSS are lower than the value of X_{TSS} = 180 g TSS m^{-3}, since a fraction of X_S in the influent would pass through membrane filters but must be included in the model component X_{TSS} since it will later adsorb onto the activated sludge. The total nitrogen (and phosphorus) in the influent may be computed with the aid of all influent concentrations multiplied with the relevant conversion factors from Tables 2.1 and 3.2.

Table 3.2 is a list of typical stoichiometric coefficients of **ASM2** and includes the factors which are required for the use of the continuity equations (see also Table 2.1). Many of the conversion factors have been estimated without performing

Table 3.1. ***Short definition of model components and typical wastewater composition (primary effluent), considering the composition of the different model components as indicated in Table 3.2.***

COD_{tot} = 260 g COD m^{-3}, TKN = 25 g N m^{-3}, TP = 6 g P m^{-3}			
Dissolved components:			
S_{O_2}	Dissolved oxygen	0	g O_2 m^{-3}
S_F	Readily biodegradable substrate	30	g COD m^{-3}
S_A	Fermentation products (acetate)	20	g COD m^{-3}
S_{NH_4}	Ammonium	16	g N m^{-3}
S_{NO_3}	Nitrate (plus nitrite)	0	g N m^{-3}
S_{PO_4}	Phosphate	3.6	g P m^{-3}
S_I	Inert, non-biodegradable organics	30	g COD m^{-3}
S_{ALK}	Bicarbonate alkalinity	5	mol HCO_3^- m^{-3}
S_{N_2}	Dinitrogen (N_2), 0.78 atm at 20 °C	15	g N m^{-3}
Particulate components:			
X_I	Inert, non-biodegradable organics	25	g COD m^{-3}
X_S	Slowly biodegradable substrate	125	g COD m^{-3}
X_H	Heterotrophic biomass	30	g COD m^{-3}
X_{PAO}	Phosphorus-accumulating organisms	0	g COD m^{-3}
X_{PP}	Stored poly-phosphate of PAO	0	g P m^{-3}
X_{PHA}	Organic storage products of PAO	0	g COD m^{-3}
X_{AUT}	Autotrophic, nitrifying biomass	0	g COD m^{-3}
X_{MeOH}	'Ferric-hydroxide' ($Fe(OH)_3$)	0	g $Fe(OH)_3$ m^{-3}
X_{FeP}	'Ferric-phosphate' ($FePO_4$)	0	g $FePO_4$ m^{-3}
X_{TSS}	Particulate material as a model component[a)]	180[a)]	g TSS m^{-3}

a) This value is larger than the TSS which may be measured analytically, since it includes the fraction of X_S, which would pass the filter in the TSS analysis. X_{TSS} may also include some inert mineral materials, which are contained in the influent but not accounted for by other components. If this is the case, then X_{TSS} in the influent may be larger than predicted from the continuity equation, which for the above values and based on the conversion factors given in Table 3.2 would result in 140 g TSS m^{-3}. Analytically measured TSS (0.45 mm) would be approximately 120 g TSS m^{-3}

Table 3.2. Definition and typical values for the stoichiometric coefficients of ASM2.

Typical conversion factors for continuity equations			
Nitrogen:			
Soluble material:			
i_{NSI}	N content of inert soluble COD S_I	0.01	g N (g COD) $^{-1}$
i_{NSF}	N content of soluble substrate S_F	0.03	g N (g COD) $^{-1}$
Particulate material:			
i_{NXI}	N content of inert particulate COD X_I	0.03	g N (g COD) $^{-1}$
i_{NXS}	N content of particulate substrate X_S	0.04	g N (g COD) $^{-1}$
i_{NBM}	N content of biomass X_H, X_{PAO}, X_{AUT}	0.07	g N (g COD) $^{-1}$
Phosphorus:			
Soluble material:			
i_{PSI}	P content of inert soluble COD S_I	0.00	g P (g COD) $^{-1}$
i_{PSF}	P content of soluble substrate S_F	0.01	g P (g COD) $^{-1}$
Particulate material:			
i_{PXI}	P content of inert particulate COD X_I	0.01	g P (g COD) $^{-1}$
i_{PXS}	P content of particulate substrate X_S	0.01	g P (g COD) $^{-1}$
i_{PBM}	P content of biomass X_H, X_{PAO}, X_{AUT}	0.02	g P (g COD) $^{-1}$
Total suspended solids:			
i_{TSSXI}	TSS to X_I ratio	0.75	g TSS (g COD) $^{-1}$
i_{TSSXS}	TSS to X_S ratio	0.75	g TSS (g COD) $^{-1}$
i_{TSSBM}	TSS to biomass ratio for X_H, X_{PAO}, X_{AUT}	0.90	g TSS (g COD) $^{-1}$
Typical stoichiometric constants			
Hydrolysis:			
f_{SI}	Fraction of inert COD in particulate substrate	0.00	g COD (g COD) $^{-1}$
Heterotrophic organisms: X_H			
Y_H	Yield coefficient	0.63	g COD (g COD) $^{-1}$
f_{XI}	Fraction of inert COD generated in biomass lysis	0.10	g COD (g COD) $^{-1}$
Phosphorus-accumulating organisms: X_{PAO}			
Y_{PAO}	Yield coefficient (biomass / PHA)	0.63	g COD (g COD) $^{-1}$
Y_{PO4}	PP requirement (S_{PO_4} release) for PHA storage	0.40	g P (g COD) $^{-1}$
Y_{PHA}	PHA requirement for PP storage	0.20	g COD (g COD) $^{-1}$
f_{XI}	Fraction of inert COD generated in biomass lysis	0.10	g COD (g COD) $^{-1}$
Nitrifying organisms: X_{AUT}			
Y_{AUT}	Yield coefficient (biomass/nitrate)	0.24	g COD (g N) $^{-1}$
f_{XI}	Fraction of inert COD generated in biomass lysis	0.10	g COD (g COD) $^{-1}$

specific experiments for their determination. These values indicate an order of magnitude. The stoichiometric coefficients are either based on previous experience with **ASM1** or they are derived from verification trials of **ASM2** relative to full-scale experience. Experience with the three yield coefficients, Y_{PAO}, Y_{PO_4} and Y_{PHA} of the PAO are still scarce.

Table 3.3 is a summary of the definitions and typical values of all kinetic parameters of the model. Again, some kinetic parameters were estimated based on the experience with **ASM1**, those relating to biological phosphorus removal are estimated based on laboratory experience and full-scale verification trials of **ASM2**. Note that saturation coefficients K_i for any specific compound may be different for different organisms (e.g. K_{O_2} may have four different values, depending on the process and organism to which it relates).

Future experience may well lead to different 'good estimates' of the parameters of the model. Since experimental results of many pilot studies have been performed without considering the requirements of model calibration, we do not currently have a sufficient basis to calibrate **ASM2** to a 'typical wastewater'.

Finally a full stoichiometric matrix, based on the proposed stoichiometric parameters in Table 3.2 is presented in Table 3.4. Table 3.4 is not meant to be a part of **ASM2** but rather it should indicate approximate values of stoichiometric coefficients ν_{ji}. Table 3.4 may be used to test computer code, which might be developed to predict stoichiometric coefficients ν_{ji} based on conversion factors and stoichiometric constants as introduced in Table 3.2.

Table 3.3. Definition and typical values for the stoichiometric coefficients of ASM2.

Temperature		20 °C	10 °C	Units
Hydrolysis:				
K_h	= Hydrolysis rate constant	3.00	2.00	d^{-1}
η_{NO_3}	= Anoxic hydrolysis reduction factor	0.60	0.60	–
η_{fe}	= Anaerobic hydrolysis reduction factor	0.10	0.10	–
K_{O_2}	= Saturation/inhibition coefficient for oxygen	0.20	0.20	$g\ O_2\ m^{-3}$
K_{NO_3}	= Saturation/inhibition coefficient for nitrate	0.50	0.50	$g\ N\ m^{-3}$
K_X	= Saturation coefficient for particulate COD	0.10	0.30	$g\ COD\ (g\ COD)^{-1}$
Heterotrophic organisms:				
μ_H	= Maximum growth rate on substrate	6.00	3.00	d^{-1}
q_{fe}	= Maximum rate for fermentation	3.00	1.50	$g\ COD\ (g\ COD)^{-1}\ d^{-1}$
η_{NO_3}	= Reduction factor for denitrification	0.80	0.80	–
b_H	= Rate constant for lysis	0.40	0.20	d^{-1}
K_{O_2}	= Saturation/inhibition coefficient for oxygen	0.20	0.20	$g\ O_2\ m^{-3}$
K_F	= Saturation coefficient for growth on S_F	4.00	4.00	$g\ COD\ m^{-3}$
K_{fe}	= Saturation coefficient for fermentation of S_F	20.00	20.00	$g\ COD\ m^{-3}$
K_A	= Saturation coefficient for S_A (acetate)	4.00	4.00	$g\ COD\ m^{-3}$
K_{NO_3}	= Saturation/inhibition coefficient for nitrate	0.50	0.50	$g\ N\ m^{-3}$
K_{NH_4}	= Saturation coefficient for ammonium (nutrient)	0.05	0.05	$g\ N\ m^{-3}$
K_P	= Saturation coefficient for phosphorus (nutrient)	0.01	0.01	$g\ P\ m^{-3}$
K_{ALK}	= Saturation coefficient for alkalinity	0.10	0.10	mole $HCO_3^-\ m^{-3}$
Phosphorus-accumulating organisms:				
q_{PHA}	= Rate constant for storage of PHA (base: X_{PP})	3.00	2.00	$g\ COD\ (g\ PAO)^{-1}\ d^{-1}$
q_{PP}	= Rate constant for storage of PP	1.50	1.00	$g\ PP\ (g\ PAO)^{-1}\ d^{-1}$
μ_{PAO}	= Maximum growth rate	1.00	0.67	d^{-1}
b_{PAO}	= Rate constant for lysis of X_{PAO}	0.20	0.10	d^{-1}
b_{PP}	= Rate constant for lysis of X_{PP}	0.20	0.10	d^{-1}
b_{PHA}	= Rate constant for lysis of X_{PHA}	0.20	0.10	d^{-1}
K_{O_2}	= Saturation coefficient for S_{O_2}	0.20	0.20	$g\ O_2\ m^{-3}$
K_A	= Saturation coefficient for S_A (acetate)	4.00	4.00	$g\ COD\ m^{-3}$
K_{NH_4}	= Saturation coefficient for ammonium	0.05	0.05	$g\ N\ m^{-3}$
K_{PS}	= Saturation coefficient for phosphorus in PP storage	0.20	0.20	$g\ P\ m^{-3}$
K_P	= Saturation coefficient for phosphorus in growth	0.01	0.01	$g\ P\ m^{-3}$
K_{ALK}	= Saturation coefficient for alkalinity	0.10	0.10	mole $HCO_3^-\ m^{-3}$
K_{PP}	= Saturation coefficient for poly-phosphate	0.01	0.01	$g\ PP\ (g\ PAO)^{-1}$
K_{MAX}	= Maximum ratio of X_{PP}/X_{PAO}	0.34	0.34	$g\ PP\ (g\ PAO)^{-1}$
K_{IPP}	= Inhibition coefficient for X_{PP} storage	0.02	0.02	$g\ PP\ (g\ PAO)^{-1}$
K_{PHA}	= Saturation coefficient for PHA	0.01	0.01	$g\ PHA\ (g\ PAO)^{-1}$
Nitrifiers:				
μ_{AUT}	= Maximum growth rate	1.00	0.35	d^{-1}
b_{AUT}	= Decay rate	0.15	0.05	d^{-1}
K_{O_2}	= Saturation coefficient for oxygen	0.50	0.50	$g\ O_2\ m^{-3}$
K_{NH_4}	= Saturation coefficient for ammonium	1.00	1.00	$g\ N\ m^{-3}$
K_{ALK}	= Saturation coefficient for alkalinity	0.50	0.50	mole $HCO_3^-\ m^{-3}$
K_P	= Saturation coefficient for phosphorus	0.01	0.01	$g\ P\ m^{-3}$
Precipitation:				
k_{PRE}	= Rate constant for P precipitation	1.0	1.0	$m^3\ (g\ Fe(OH)_3)^{-1}\ d^{-1}$
k_{RED}	= Rate constant for redissolution	0.6	0.6	d^{-1}
K_{ALK}	= Saturation coefficient for alkalinity	0.50	0.50	mole $HCO_3^-\ m^{-3}$

***Table 3.4.** An example of a stoichiometric matrix for* **ASM2** *for soluble and particulate components and for precipitation processes. The absolute values of the stoichiometric coefficients are based on the typical stoichiometric parameters introduced in Table 3.2. These values are not the* **ASM2** *but rather a typical application of the model.*

Stoichiometric matrix for dissolved components

Process:	S_{O_2}	S_F	S_A	S_{NH_4}	S_{NO_3}	S_{PO_4}	S_I	S_{ALK}	S_{N_2}
1 Aerobic hydrolysis		1.00		0.01			0.00	0.001	
2 Anoxic hydrolysis		1.00		0.01			0.00	0.001	
3 Anaerobic hydrolysis		1.00		0.01			0.00	0.001	
Heterotrophic organisms: X_H									
4 Growth on S_F	-0.59	-1.59		-0.022		-0.004		-0.001	
5 Growth on S_A	-0.59		-1.59	-0.070		-0.02		0.021	
6 Denitrification with S_F		-1.59		-0.022	-0.21	-0.004		0.014	0.21
7 Denitrification with S_A			-1.59	-0.07	-0.21	-0.02		0.036	0.21
8 Fermentation		-1	1.00	0.03		0.01		0.014	
9 Lysis				0.031		0.01		0.002	
Phosphorus-accumulating organisms (PAO): X_{PAO}									
10 Storage of X_{PHA}			-1			0.40		-0.004	
11 Storage of X_{PP}	-0.20					-1		0.048	
12 Aerobic growth	-0.60			-0.07		-0.02		-0.004	
13 Lysis of X_{PAO}				0.031		0.01		0.002	
14 Lysis of X_{PP}						1.00		-0.048	
15 Lysis of X_{PHA}			1.00					-0.016	
Nitrifying organisms (autotrophic organisms): X_{AUT}									
16 Growth	-18.0			-4.24	4.17	-0.02		-0.600	
17 Lysis				0.031		0.01		0.002	
Simultaneous precipitation of phosphorus with ferric hydroxide ($Fe(OH)_3$):									
18 Precipitation						-1			0.048
19 Redissolution						1			-0.048

Stoichiometric matrix for particulate components

Process:	X_I	X_S	X_H	X_{PAO}	X_{PP}	X_{PHA}	X_{AUT}	X_{TSS}	X_{MeOH}	X_{MeP}
1 Aerobic hydrolysis		-1						-0.75		
2 Anoxic hydrolysis		-1						-0.75		
3 Anaerobic hydrolysis		-1						-0.75		
Heterotrophic organisms: X_H										
4 Growth on S_F			1					0.90		
5 Growth on S_A			1					0.90		
6 Denitrification with S_F			1					0.90		
7 Denitrification with S_A			1					0.90		
8 Fermentation										
9 Lysis	0.10	0.90	-1.00					-0.15		
Phosphorus-accumulating organisms (PAO): X_{PAO}										
10 Storage of X_{PHA}					-0.40	1.00		-0.69		
11 Storage of X_{PP}					1.00	-0.20		3.11		
12 Aerobic growth				1		-1.60		-0.06		
13 Lysis of X_{PAO}	0.10	0.90		-1				-0.15		
14 Lysis of X_{PP}					-1			-3.23		
15 Lysis of X_{PHA}						-1		-0.60		
Nitrifying organisms (autotrophic organisms): X_{AUT}										
16 Growth							1	0.90		
17 Lysis	0.10	0.90					-1	-0.15		
Simultaneous precipitation of phosphorus with ferric hydroxide ($Fe(OH)_3$):										
18 Precipitation								1.42	-3.45	4.87
19 Redissolution								-1.42	3.45	-4.87

4. Wastewater characterization for activated sludge processes

The **Activated Sludge Model No. 2** (**ASM2**) is a working tool for nutrient removal treatment plants. The quality of the model predictions will depend on the quality of the wastewater characterization and on the calibration of the model.

A detailed knowledge of the influent to a wastewater treatment system will, in itself, allow for a good prediction of the performance of the system. The wastewater composition influences the actual system performance, to a degree similar to that of the system design. The characterization of a given wastewater can be made by a more or a less detailed procedure. The more detailed the characterization, the more reliable the results obtained from the modelling effort will be.

It is thus the planned use of the modelling results that determines the degree of complexity needed in the characterization. If the results are to be used for design purposes, then detailed characterization is needed. If the model is used for teaching purposes, then much less complexity is needed in the characterization.

4.1 Variations in wastewater composition

ASM2 has been developed for systems treating municipal wastewater, with only minor contributions or impact from industrial discharges. This means that industrial effluents do not cause major changes in the overall composition of the 'domestic' wastewater.

The wastewater entering a wastewater treatment plant will have its detailed composition determined by three factors:

- Wastewater input to the sewer
- Sewer system type (separate/combined)
- Transformation processes in the sewer

The wastewater input to the sewer can vary due to rain, industry and the habits of the population connected to the sewer system. The variations in input lead to variations in concentrations of the various components. For municipal wastewater with only a minor contribution from industrial effluents, the concentration of the components, and the ratios between the various components, are not severely influenced by the industrial discharges.

The transformation processes in the sewer depend on temperature, transport time, and oxygen supply to the wastewater being transported. Apart from seasonal variations in temperature, the transformation processes are not sensitive to the variation in wastewater concentrations found in the sewer. This means that the change in the ratios between the various components, caused by the transformation processes in the sewer, is relatively constant with time. The result with respect to the various fractions in the wastewater entering the treatment plant is that they do not vary very much with time, although the actual influent concentrations vary significantly with time and day.

4.2 Characterization of wastewater

ASM2 can be used for any type of municipal wastewater: raw wastewater, primary settled or preprecipitated wastewater. Pretreatment of wastewater affects the distribution among the fractions in the wastewater significantly. **ASM2** does not include pretreatment as a model feature. The input to the model must be based on the characteristics of the input to the biological tanks in the treatment plant.

4.2.1 Organic fractions in municipal wastewater

The methods for characterizing the organic fractions are still under development, and are not standardized. It is important to be aware of the close relationship between characterization of wastewater and the models and constants used. A method of characterization may well prove effective in a certain modelling context, and be of no use in another context. One example of this is the measurement of soluble inert material (see below).

The total organic matter content in wastewater can be measured as COD, C_{TCOD}. This can be subdivided, depending on the complexity and the use of the model. Figure 4.1 shows the COD fractions used in the characterization of wastewater (and mixed liquor). The typical relative distribution for primary settled wastewater is shown. The COD measurements must be made with dichromate, not permanganate, to ensure correct mass balances in the calculations. Table 4.1 gives typical ranges for the COD fractions used in **ASM2**.

The intervals given for the particulate components cover raw and primary settled wastewater. For preprecipitated wastewater, the particulate concentrations are smaller than the minimum values given in Table 4.1.

The broad range given for the components covers situations with separate sewers, low infiltration and limited water resources (high wastewater concentrations, as in South Africa); and combined sewers, high infiltration and plentiful water resources (low wastewater concentrations, as in Scandinavia and Switzerland).

According to Figure 4.1 and Table 4.1, the total COD in the model includes the following components

$$C_{TCOD} = S_A + S_F + S_I + X_I + X_S + X_H + X_{PAO} + X_{PHA} + X_{AUT} \qquad (4.1)$$

Not all the components shown in Equation 4.1 are of equal importance. The biomass fractions in the influent are needed in order to explain the bloom and the wash-out of certain groups of microorganisms. If seeding from the influent is very high, then the processes may occur even in high-load situations, where wash-out would be expected. For heterotrophs (and the fraction of these that can denitrify), the growth rate is so high that wash-out never occurs in practice. They need not be considered in the influent, because the initial concentrations used for most calculations

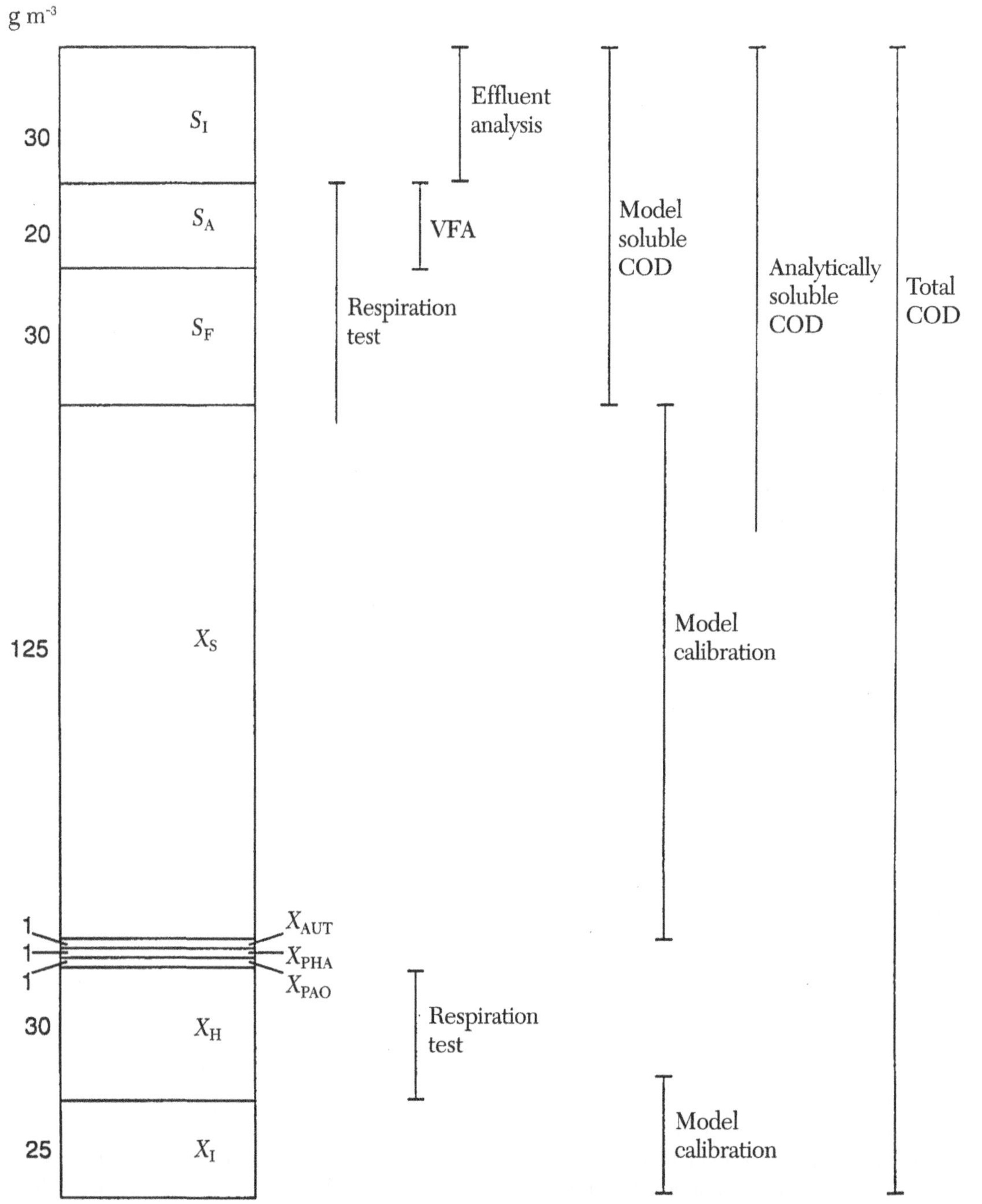

Figure 4.1. COD fractionation in **ASM2**. *The column shows a typical distribution of COD in primary effluent from municipal wastewater treatment. Various analytical techniques can measure parts of the COD as indicated on the figure.* X_I *is obtained through modelling using sludge production.* X_S *is found by modelling using oxygen uptake rate/nitrogen uptake rate (OUR/NUR) test results.*

Table 4.1. Municipal wastewater components in ASM2.

Symbol	Component	Typical range	Unit
Model dissolved components:			
S_{O_2}	Dissolved oxygen	0.0–0.5	g O_2 m^{-3}
S_F	Readily (fermentable) biodegradable substrate	20–250	g COD m^{-3}
S_A	Volatile acids/fermentation products (acetate)	10–60	g COD m^{-3}
S_{NH4}	Ammonium nitrogen	10–100	g N m^{-3}
S_{NO3}	Nitrate plus nitrite nitrogen	0–1	g N m^{-3}
S_{PO4}	Phosphate phosphorus	2–20	g P m^{-3}
S_I	Inert, non-biodegradable organics	20–100	g COD m^{-3}
Model 'particulate' components[a)]:			
X_I	Inert, non-biodegradable organics	30–150	g COD m^{-3}
X_S	Slowly biodegradable substrate[a)]	80–600	g COD m^{-3}
X_H	Heterotrophic biomass	20–120	g COD m^{-3}
X_{PAO}	Phosphorus-accumulating organisms	0–1	g COD m^{-3}
X_{PP}	Stored poly-phosphate phosphorus in PAO	0–0.5	g P m^{-3}
X_{PHA}	Stored poly-hydroxy-alkanoate	0–1	g COD m^{-3}
X_{AUT}	Autotrophic, nitrifying biomass	0–1	g COD m^{-3}

a) Some of which are analytically soluble.

always assume the presence of these organisms.

The phosphotrophs (the phosphate-accumulating biomass) and the autotrophs should be included in order to account for the development and the wash-out phenomena related to these two biomass groups. These two groups can be washed out of high-loaded treatment plants. The concentration of autotrophs in the influent is in most cases very small, and this is also believed to be the case for the phosphate-accumulating biomass, X_{PAO}. Note that X_{PAO} does not include stored poly-hydroxy-alkanoates, X_{PHA}, which are considered a separate compound. In a similar manner, the poly-phosphate stored in the phosphotrophs is considered a separate compound. This means that the contribution of the phosphotrophs to the total suspended solids comes from three fractions (X_{PAO}, X_{PHA} and X_{PP}).

Stored poly-hydroxy-alkanoate, X_{PHA}, is close to zero in raw wastewater. This means that the total COD fractionation in many cases can be simplified to

$$C_{TCOD} = S_A + S_F + S_I + X_I + X_S + X_H \qquad (4.2)$$

or, in cases where the heterotrophic biomass is negligible or included in the slowly degradable suspended organics, X_S, to

$$C_{TCOD} = S_A + S_F + S_I + X_I + X_S \qquad (4.3)$$

The inclusion of X_H in X_S does not affect the modelling significantly, but it affects the value of the yield coefficient Y_H (a smaller yield coefficient must be chosen).

Based on the various simplifications, Equations 4.1 to 4.3 can be used for calculation of those parts of the organic components that cannot be measured directly.

For municipal wastewater, the fractions of the various organic components will normally be within a limited range. For a specific wastewater, the daily and seasonal variations seem to be within a relatively narrow range in most cases. In Table 4.2 some typical ranges are given. Pretreatment can change these ranges. For example primary sludge acidification can increase the acetic acid concentration, and thus S_A, considerably.

4.2.2 Nitrogen fractions in municipal wastewater

In general there is no need to characterize the nitrogen fractions in as much detail as for organic matter. One reason for this is that the major part

Table 4.2. Typical ranges for the organic fractions of municipal wastewater, primary effluent.

Symbol	Component	Typical fraction of total COD in %
S_F	Readily (fermentable) biodegradable substrate	10–20
S_A	Volatile acids (acetate)	2–10
S_I	Inert, non-biodegradable organics	5–10
X_I	Inert, non-biodegradable organics	10–15
X_S	Slowly biodegradable substrate	30–60
X_H	Heterotrophic biomass	5–15
X_{PAO}	Phosphorus-accumulating organisms	0–1
X_{PHA}	Stored poly-hydroxy-alkanoate	0–1
X_{AUT}	Autotrophic, nitrifying biomass	0–1

Table 4.3. Fractions of nitrogen and phosphorus in organic matter in municipal wastewater.

Symbol	Component	Typical range N	Typical range P	Unit
S_F	Readily (fermentable) biodegradable substrate	2–4	1–1.5	% of COD
S_A	Volatile acids (acetate)	0	0	% of COD
S_I	Inert, non-biodegradable organics	1–2	0.2–0.8	% of COD
X_I	Inert, non-biodegradable organics	0.5–1	0.5–1	% of COD
X_S	Slowly biodegradable substrate	2–4	1–1.5	% of COD
X_H	Heterotrophic biomass	5–7	1–2	% of COD
X_{PAO}	Phosphorus-accumulating organisms	5–7	1–2	% of COD
X_{PHA}	Stored poly-hydroxy-alkanoate	0	0	% of COD
X_{AUT}	Autotrophic, nitrifying biomass	5–7	1–2	% of COD

of the nitrogen in wastewater is present as ammonia, which has no coupling to the organic components. For the remaining part of the nitrogen, most of which is coupled to the organic components, it is sufficient to use fixed nitrogen fractions for the various COD components, as shown in Table 4.3.

Several nitrogen fractions can be easily determined by the use of standard chemical analysis as shown in Figure 4.2.

The total nitrogen concentration in municipal wastewater, C_{TN}, can be characterized as

$$C_{TN} = C_{TKN} + S_{NO_3} = X_{TKN} + S_{TKN} + S_{NO_3} \quad (4.4)$$

where

C_{TKN} is total Kjeldahl nitrogen,

X_{TKN} is particulate Kjeldahl nitrogen, and

S_{TKN} is soluble Kjeldahl nitrogen.

From Table 4.3 it is seen that all organic particulate fractions contain nitrogen, except for X_{PHA}, stored poly-hydroxy-alkanoate. X_{TKN} is the sum of nitrogen bound to all the other organic particulate fractions, as given in Table 4.3.

$$X_{TKN} = (X_I \cdot i_{NX_I}) + (X_S \cdot i_{NX_S}) + (X_H + X_{PAO} + X_{AUT})\, i_{NBM}. \quad (4.5)$$

The soluble Kjeldahl nitrogen is dominated by ammonium-nitrogen, S_{NH_4}

$$S_{TKN} = S_{NH_4} + (S_F \cdot i_{NS_F}) + (S_I \cdot i_{NS_I}) \quad (4.6)$$

The soluble inert nitrogen fraction, $S_I \cdot i_{NS_I}$, is important in relation to very strict effluent criteria. Normally the nitrogen fraction i_{NS_I} is small, as seen from Table 4.3, and the concentrations found in raw municipal wastewater are in the range 0.5–1 g N m^{-3}. Elevated values might be a result of industrial discharges or high strength wastes.

4.2.3 Phosphorus fractions in municipal wastewater

In general there is no need to characterize the phosphorus fractions in as much detail as for organic matter. For many modelling purposes, it is sufficient to couple a fixed phosphorus fraction to the various COD fractions, as in Table 4.3. Figure 4.3 shows a typical distribution of phosphorus in primary effluent.

The total phosphorus concentration in raw municipal wastewater can be divided into these fractions:

$$C_{TP} = X_{TP} + S_{TP} \quad (4.7)$$

where

X_{TP} is particulate phosphorus, and

S_{TP} is soluble phosphorus.

The particulate phosphorus, X_{TP}, includes inorganic phosphorus (expressed as 'ferric phosphate', X_{FeP}) and organic phosphorus.

$$X_{TP} = (0.205 \cdot X_{FeP}) + X_{PP} + (X_S \cdot i_{PX_S}) + (X_I \cdot i_{PX_I}) + (X_H + X_{PAO} + X_{AUT}) \cdot i_{PBM} \quad (4.8)$$

In municipal wastewater the stored poly-phosphate concentration, X_{PP}, is close to zero and, for many wastewaters, the same is true for the metal phosphate concentration, X_{FeP}. The contribution to the particulate phosphorus concentration from autotrophs and phosphotrophs is in most cases negligible. This reduces Equation 4.8 to:

$$X_{TP} = (X_S \cdot i_{PX_S}) + (X_I \cdot i_{PX_I}) \quad (4.9)$$

The soluble phosphorus includes

$$S_{TP} = S_{PO_4} + (S_F \cdot i_{PS_F}) + (S_I \cdot i_{PS_I}) \quad (4.10)$$

The soluble inorganic phosphorus concentration, S_{PO_4}, which for wastewater will consist of ortho-phosphate and poly-phosphates is, in **ASM2**, considered as ortho-phosphate only. In municipal wastewater the concentration of soluble organic phosphorus is small compared with the inorganic ortho-phosphate concentration. Thus for municipal wastewater the soluble phosphorus can be approximated by:

$$S_{TP} = S_{PO_4} \quad (4.11)$$

4.3 Routine analysis of model components

A number of the components in **ASM2** can be analysed by use of routine chemical analysis. These components are mentioned below.

S_A, volatile acids/fermentation products. The volatile acids/fermentation products in municipal wastewater are dominated by acetic acid, which normally accounts for 60–80% of the COD in this fraction. The concentration varies considerably from place to place, mainly as a result of the changes the wastewater undergoes during transport in the sewer.

S_A, can be estimated by respiration tests, either with oxygen (OUR) or with nitrate (NUR). The respiration tests do not seem to be very accurate. The volatile acid fraction of S_A, can be measured directly by gas chromatography.

C_{TN}, total nitrogen. Total nitrogen in raw municipal wastewater is dominated by reduced nitrogen, either in the form of ammonia or as amino groups in organic substances. Ammonia is the dominating reduced nitrogen component, and accounts normally for 60–70% of the Total Kjeldahl Nitrogen. Small concentrations of nitrate or nitrite can be found, often in the range 0–1 g m^{-3}. Total nitrogen can be calculated from Equation 4.4:

$$C_{TN} = C_{TBN} + S_{NO3}$$

C_{TKN}, Total Kjeldahl Nitrogen. Total Kjeldahl Nitrogen (TKN) in raw municipal wastewater includes organic reduced nitrogen (amino groups) and ammonia. Most of the TKN is of physiological origin. It is measured by traditional chemical analysis (Kjeldahl analysis).

S_{NH4}, ammonia-nitrogen. Ammonia in raw municipal wastewater has its primary origin in urea, which is very quickly hydrolysed and seldom found – or looked for – at the wastewater treatment plant. Ammonia can be measured by traditional chemical analytical methods.

S_{NO3}, nitrate- and nitrite-nitrogen. Oxidized nitrogen in raw municipal wastewater can have its origin in infiltration water or, if occurring in

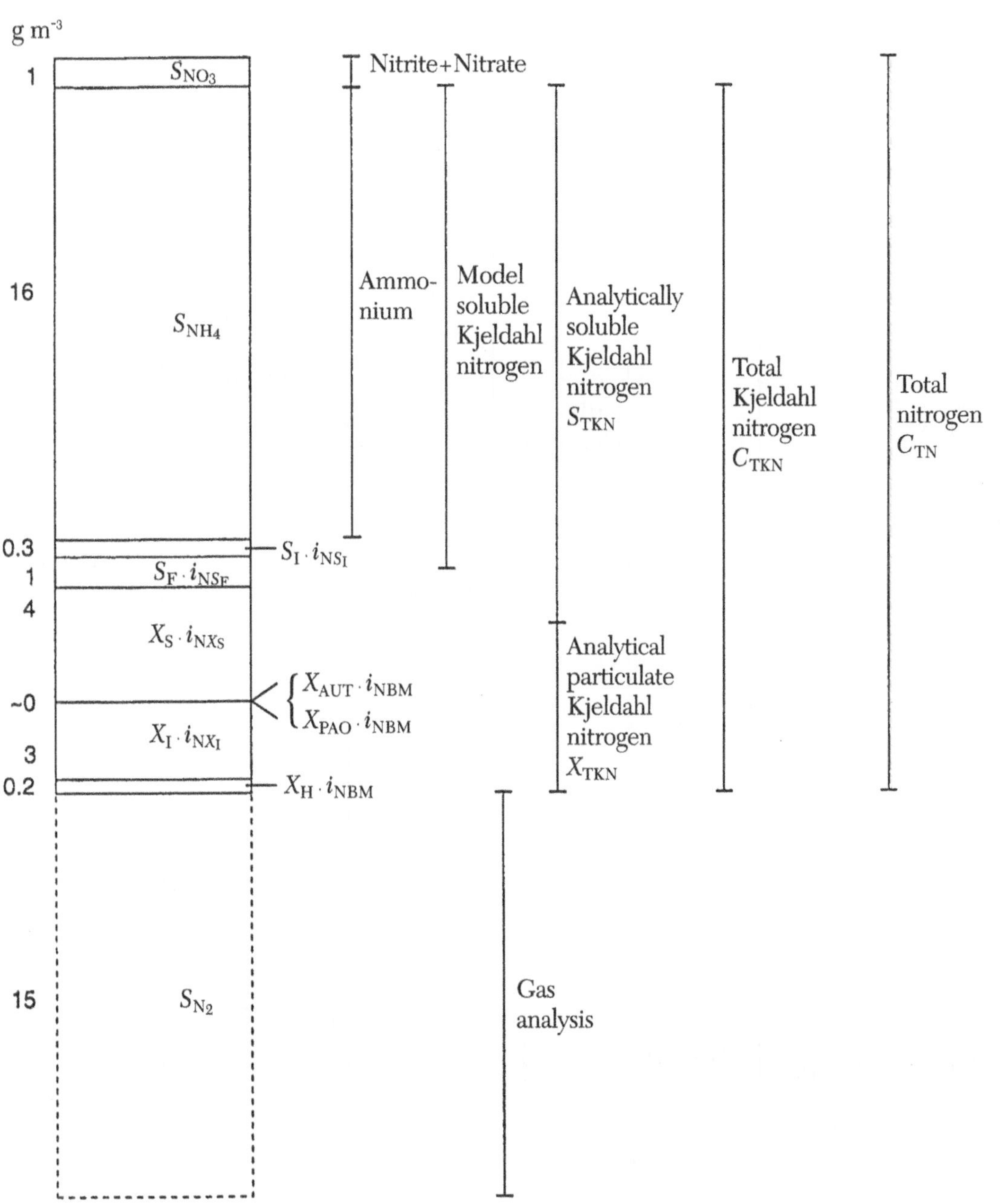

*Figure 4.2. Nitrogen fractionation in **ASM2**. The column shows a typical distribution of nitrogen in primary effluent from municipal wastewater treatment. Various analytical techniques can measure parts of the nitrogen as indicated on the figure.*

elevated concentrations, drinking water and industrial effluents. Nitrite and nitrate can be measured by traditional chemical analytical methods, either separately or combined.

C_{TP}, total phosphorus. Total phosphorus in raw municipal wastewater is in the phosphate form, either as inorganic or as organic bound phosphorus. The major part of organic bound phosphorus is of physiological origin. The total concentration of phosphorus and the various fractions present in raw wastewater, are heavily influenced by the use of phosphorus in detergents. The phosphorus in detergents will, in raw wastewater, occur as soluble poly-phosphates or (after hydrolysis) as ortho-phosphate. Phosphorus from detergents can account for up to 50% of the total phosphorus concentration in raw wastewater. Total phosphorus is measured by traditional chemical analytical methods.

S_{PO_4}, ortho-phosphate. Much of the ortho-phosphate present in municipal wastewater has its origin in detergents and other household chemicals. The poly-phosphate in detergents hydrolyses slowly to ortho-phosphate, and part of this process occurs in the sewer. Ortho-phosphate is measured by traditional chemical analytical methods.

4.4 Soluble analysis as a tool in characterization

Soluble material does not have a universal definition in wastewater when used in a modelling context. It is coupled to the whole set of constants and components used in an actual model. In traditional wastewater analysis, the components are often split into two main fractions, the soluble and the particulate (suspended). Filters with a pore size of 0.45–1.8 μm are normally used for separation of the two main fractions in the wastewater.

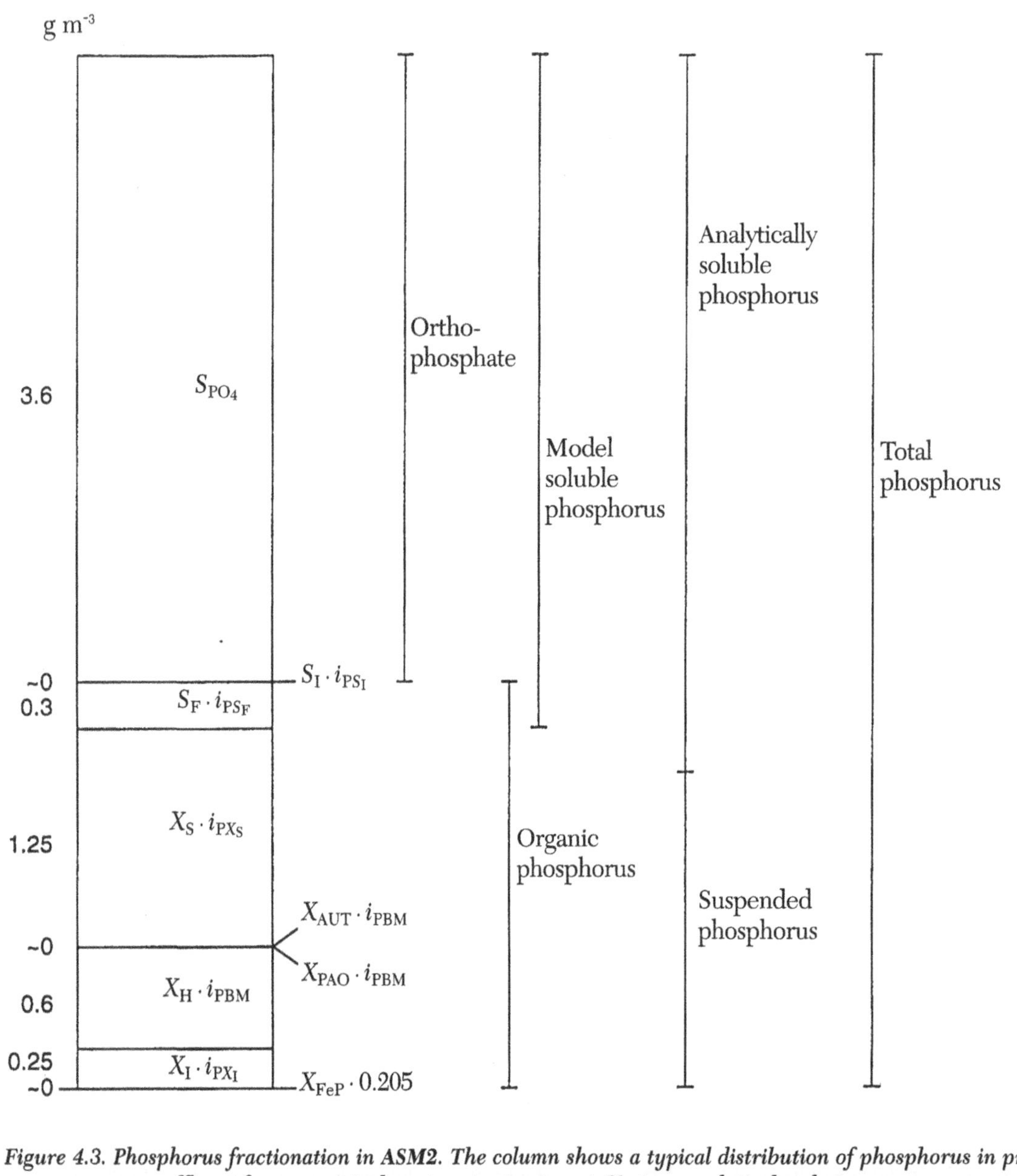

Figure 4.3. Phosphorus fractionation in ASM2. The column shows a typical distribution of phosphorus in primary effluent from municipal wastewater treatment. Various analytical techniques can measure parts of the phosphorus as indicated on the figure.

For some purposes like chemical precipitation, a third fraction, colloidal matter, can be advantageous.

For modelling purposes using **ASM2**, we cannot analytically distinguish particulate and dissolved components in wastewater by filtration. Tests like those described in Section 4.6 for S_F and X_S have to be used. However, the use of soluble analysis will improve the wastewater characterization by imposing some limitations on the results from various analysis and model calibrations. Examples of relationships between analytical soluble COD, S_{TCOD}, and the components in **ASM2** are

$$S_{TCOD}\,(\text{analytical}) > S_{TCOD}\,(\text{model}) = S_I + S_A + S_F \quad (4.12)$$

$$S_{TCOD}\,(\text{analytical}) < S_I + S_A + S_F + X_S \quad (4.13)$$

$$C_{TCOD}(\text{analytical}) - S_{TCOD}(\text{analytical}) = X_{TCOD}\,(\text{analytical}) < X_{TCOD}(\text{ model}) = X_S + X_I + X_H + X_{PAO} + X_{PHA} + X_{AUT} \quad (4.14)$$

For analytical soluble Kjeldahl nitrogen:

$$S_{TKN}\,(\text{analytical}) > S_{NH_4}\,(\text{analytical}) = S_{NH_4}(\text{model}) \quad (4.15)$$

For analytical soluble total phosphorus:

$$S_{TP}\,(\text{analytical}) > S_{PO_4}(\text{analytical}) < S_{PO_4}(\text{model}) \quad (4.16)$$

In mixed liquor, the measured soluble COD fractions are the correct ones, because the part of the slowly biodegradable substrate, X_S, which is soluble in wastewater, will be adsorbed onto the activated sludge particles.

4.5 Model components without standardized analytical procedures

Many model components cannot be measured directly or precisely. In some cases the analytical procedures are not yet standardized, due to a need for further understanding and development. This is the case for stored poly-phosphate, X_{PP}, and for a long list of COD components such as S_F, S_I, X_I, X_S, X_H, X_{AUT}, X_{PAO} and X_{PHA}.

There is a need for further development of methods for measuring all the above mentioned components. For those components which are used in **ASM1**, S_I, X_I and X_S, various measurement techniques have been developed. These can in many cases give reasonable results, but the methods still need further development and refinement.

The methods for estimating the three biomass fractions, X_H, X_{AUT} and X_{PAO}, are all based on bulk activity measurements, not on microbiological methods. In wastewater, the activities of the autotrophs and the phosphotrophs are so small that they cannot be measured. In mixed liquor it is possible to make activity measurements and from these calculate the biomass fraction present. The estimation of X_{PAO} is the most questionable, as the phosphotroph behaviour is not well known.

In municipal wastewater X_{AUT}, X_{PAO} and X_{PHA} can be assumed to be close to zero in most cases. In mixed liquor this is definitely not true. Thus, although usually required for wastewater characterization, methods are also needed for mixed liquor characterization in order to calibrate **ASM2** for process optimization.

For nitrogen and phosphorus compounds there are no significant analytical problems, except for stored poly-phosphate in mixed liquor from activated sludge treatment with biological phosphorus uptake.

4.6 Present status for measurement/ estimation of problematic components

S_F, readily (fermentable) biodegradable substrate. The readily biodegradable substrate is composed of small molecules that can be metabolized directly, or quickly fermented/hydrolysed before being metabolized. The compounds can be soluble proteins and carbohydrates and similar easily degradable compounds.

Determination of this fraction can be made biologically, chemically, physically or physico-chemically.

In **ASM2** the soluble substrate, S_S, has been split into two parts, S_A, volatile acids/fermentation products, and S_F, readily (fermentable) biodegradable substrate. This means that the soluble substrate, S_S, known from **ASM1** no longer exists, except as the sum of these two new compounds.

Biological determination is believed to be the safest at present. It is based on a respiration test with either oxygen (OUR) or nitrate (NUR). As for all other modelling activities, the safest procedure is to use a method similar to the process in the treatment plant to be simulated, as well as the sludge from the treatment plant. If the first oxidation step in the plant is with oxygen, then OUR will give the best estimates of S_F in that process. If denitrification is the first oxidation step, then NUR will give the safest estimate. Observe that the two measurements will not necessarily give exactly similar results. The OUR test must be inhibited or compensated for nitrification. From the integrated oxygen or nitrate consumption, the concentration of readily biodegradable substrate can be calculated, assuming a yield factor (Ekama *et al*., 1986; Kristensen *et al*., 1992). Alternatively, curve-fitting using a model can be used (Kappeler and Gujer, 1992). If the acetic acid concentration, S_A, has been measured then S_F can be found as

$$S_F = S_S - S_A \quad (4.17)$$

The chemical determination of soluble proteins and carbohydrates is based on traditional chemical analysis. The sum of these two parameters can be used as an estimate for S_F.

A physical determination can be made by ultrafiltration or gelfiltration, using cut-off values around 1000 daltons. The COD concentration in the filtrate, $S_{filtrate}$ is

$$S_{filtrate} = S_A + S_F + S_{I,f} \quad (4.18)$$

where $S_{I,f}$ is the inert COD in the filtrate (Dold *et al.*, 1986). The $S_{I,f}$ fraction is small, less than 10% of the filtrate for municipal wastewater (Henze, 1992). In many cases $S_{I,f}$ is close to S_I.

The physico-chemical determination of S_F can be made by chemical precipitation, followed by COD measurements. The soluble fraction after precipitation, S_{precip}, has a similar composition to $S_{filtrate}$

$$S_{precip} = S_A + S_F + S_{I,f} \quad (4.19)$$

The precipitation can be made with zinc sulphate (Mamais *et al.*, 1993) or other precipitants like polymerized aluminium chloride (Henze and Harremoës, 1992).

S_I, inert non-biodegradable organics. This soluble fraction consists of molecules of varying size. There is a small fraction of low molecular weight which can be found in the filtrate from ultra-filtration. Medium-sized inert molecules are infrequent, which means that the major part of soluble inert organics are large molecules (molecular weight above 100,000). The inert soluble organics contain small amounts of nitrogen only (see Table 4.3).

The determination of soluble inert organics in the wastewater depends on the handling of these substances in the model used. In **ASM2** there is no generation of soluble inerts in the processes. All soluble inerts are assumed to be present in the influent. This is of course not strictly correct (Orhon *et al.*, 1989), but for municipal wastewater it is close to reality.

The determination of soluble inerts in relation to **ASM2** can be determined from a long-term soluble BOD test. The soluble COD remaining after 20 days of oxidation can be regarded as equivalent to S_I (Ekama *et al.*, 1986).

The soluble inert COD can also be determined from continuous lab- or pilot-scale experiments with high solids retention time. In this case the major part (90–95%) of the soluble COD in the effluent will be inert and thus represent S_I.

Other less established methods using various batch tests (combinations of total samples and soluble samples), aerated for prolonged periods, also allow for estimating S_I (Lesouef *et al.*, 1992).

X_S, slowly degradable organic substrate. This component is normally found as the difference between the total COD and the other fractions:

$$X_S = C_{TCOD} - S_A - S_F - S_I - X_H - X_{PAO} - X_{AUT} - X_I \quad (4.20)$$

In municipal wastewater, Equation 4.20 simplifies to

$$X_S = C_{TCOD} - S_A - S_F - S_I - X_H - X_I \quad (4.21)$$

In some modelling cases, the slowly biodegradable organics in the model include the biomass fractions; in other cases these fractions may be negligible.

The slowly degradable organic substrate can also be estimated based on batch or continuous experiments, but these estimates will not be very accurate (Ekama *et al.*, 1986; Kappeler and Gujer, 1992).

X_H, heterotrophic biomass. The heterotrophic biomass present in raw wastewater will often inoculate the activated sludge process significantly, especially in cases where primary settling is not used. If the heterotrophic biomass in the raw wastewater has kinetic characteristics that deviate from those normally found in activated sludge, this will influence process kinetics. A negative influence can be due to a low growth rate or due to a high fraction of filamentous bacteria (Gujer and Kappeler, 1992). The heterotrophic biomass can ideally be subdivided into several fractions. The three main fractions in relation to biological nitrogen and phosphorus removal are the phosphotrophs, the denitrifiers and the aerobic heterotrophs. The last group can neither denitrify nor accumulate poly-phosphate. In practice these groups overlap, which makes the modelling process more complicated.

In **ASM2** the heterotrophic biomass structure has been simplified. It includes two separate fractions, the phosphotrophs, X_{PAO}, and the 'heterotrophs', X_H (denitrifiers and non-denitrifiers). Experience shows that the model with this simplification can predict the behaviour of biological nitrogen and phosphorus removal treatment plants reasonably well.

The total heterotrophic biomass ($X_H + X_{PAO}$) can be estimated from a batch experiment. Using an optimized mixture of centrifuged wastewater and mixed liquor will allow for curve-fitting, and determination of not only the heterotrophic biomass, but also the readily biodegradable substrate. The optimal mixture is approximately 90% wastewater and 10% mixed liquor. The mixture must be such that the biomass growth curve and oxygen utilization rate curve give a response that is dynamic enough to allow for a good curve-fitting for both parameters in the wastewater, and in addition the heterotrophic growth rate (Kappeler and Gujer, 1992).

The heterotrophic biomass can also be measured from a batch experiment with raw wastewater oxygen respiration. It is important that a sufficient supply of volatile acids, S_A, is present during the part of the experiment used for the biomass calculation. If the wastewater does not contain enough acetate or generate enough through fermentation, then acetate can be added to the batch (Kristensen *et al.*, 1992). The heterotrophic growth rates obtained from some of these batch tests are rather high, and probably valid for high-loaded activated sludge processes only. The reason being that the conditions in the batch tests are such that fast-growing heterotrophs proliferate. These fast growers might not develop in a low-loaded process. Detailed analysis of the growth curves obtained in batch tests seems to allow for more accurate estimation of the maximum specific growth rates for heterotrophs (Novak *et al.*, 1994).

X_{PAO}, phosphate-accumulating biomass, phosphotrophs. The phosphate-accumulating biomass cannot at present be measured reliably in raw wastewater. It is believed that the concentration of these organisms is low, because raw wastewater has almost zero phosphate-accumulation capacity.

X_I, inert suspended organics. This fraction can only be found based on continuous lab- or pilot-scale experiments. The best estimate of X_I is obtained by comparing measured and predicted sludge production. For computer programs based on a defined solids retention time, X_I can be found by calibration of the mixed liquor concentration measured in the activated sludge tanks.

X_{PP}, stored poly-phosphate in poly-phosphate-accumulating biomass. This fraction is normally close to zero in raw municipal wastewater. Stored poly-phosphate is measured by acid extraction from the suspended phosphorus (Mino *et al.*, 1984).

5. Calibration of the Activated Sludge Model No. 2

The calibration of models can be undertaken at various levels and with various resources. A calibration is always based on results from experiments performed with the actual wastewater, or the process layout to be studied, or preferably both factors. Such experiments must be properly designed in order to give optimal information for the calibration. The experiments can be performed in the full-scale plant, if the model has to be used for optimization, or in a pilot plant. In all cases, the design of the experiments should be carried out using a model to predict the experimental results, in order to ensure that the experimental conditions are optimal.

An experiment, or a series of similar experiments, will normally only allow for calibration of a few of the many constants in the model. The more measurements and experimental results available, the more constants can be calibrated, and the more reliable the calibration and the results obtained with the model will be.

The calibration of the **Activated Sludge Model No. 2 (ASM2)**, must be based on a mixture of experiments and computer simulations with the model.

In order to fit the model to experimental data, only a few model parameters usually need to be changed. The changes made should be based on the following principles:

1. Most parameters should not be changed during calibration, because they do not seem to vary significantly from case to case. Some of these relatively stable parameters are listed in Table 5.1.
2. If the experiments used for the calibration are such that the results are non-sensitive to variations in a given parameter, then that parameter should not be changed. If, for some reason, changing the parameter is unavoidable, it should only be changed in a logical direction based on some experience. All other changes are dangerous, because they could distort the model too far away from reality.
3. Only one parameter should be changed at a time. Often the parameters are highly interactive so that it is difficult to assess simultaneous changes in two or three parameters. For parameter pairs which interact, only the parameter with the biggest relative influence should be changed. This means that in the case of, for example, growth and decay rate calibration, only the growth rate should be changed.
4. Due to the highly interactive nature of some of the parameters, it is often difficult to isolate which parameter should be changed. To resolve this problem, experimental conditions should be such that the effect of the parameter of interest dominates the response.

There are mathematical calibration techniques for models, which could be applied to **ASM2**. These techniques can, however, only handle three to four parameters at a time. Thus this will not solve the general problem of this model with its numerous parameters. Further research is needed for improving calibration techniques for **ASM2**.

The limited experience available with calibration of **ASM2** shows that it can be calibrated based on a logical step-wise procedure, and by changing just a few of the many constants. The

Table 5.1 Model parameters that are relatively constant from case to case

Parameter	Name	Typical value, 20 °C	Unit
Y_H	Heterotrophic yield on S_F and S_A	0.63	$g\ COD\ (g\ COD)^{-1}$
Y_{AUT}	Autotrophic yield on nitrate produced	0.24	$g\ COD\ (g\ N)^{-1}$
μ_H	Heterotrophic growth rate on substrates S_F and S_A	6	d^{-1}
K_{O_2}	Heterotrophic saturation coefficient for oxygen	0.2	$g\ O_2\ m^{-3}$
K_F and K_A	Heterotrophic saturation coefficient for substrates S_F and S_A	4	$g\ COD\ m^{-3}$
K_{NO_3}	Heterotrophic saturation coefficient for nitrate-nitrogen	0.5	$g\ N\ m^{-3}$
K_{O_2}	Autotrophic saturation coefficient for oxygen	0.5	$g\ O_2\ m^{-3}$
K_{NH_4}	Autotrophic saturation coefficient for ammonium-nitrogen	1.0	$g\ N\ m^{-3}$

apparent high degree of freedom does not exist in practice, due to the coupling of the many processes and the use of mass balances.

It must be emphasized that a reasonable calibration will only be possible with a detailed understanding of the principles of the model. Without this, calibration will be extremely time-consuming and, in many cases, lead to combinations of parameter values that will result in models which are not suitable for extrapolation (or even interpolation).

5.1 Calibration levels

There is no general method of calibration which can be used in all cases. This is discussed below, and demonstrated in Chapter 6 where different calibration approaches are presented. Calibration of models can be made at different levels of complexity, based on the amount of data available and the planned use of the simulation results.

The best calibration is obtained when the data used for the calibration are of a type similar to those data which are required from the simulations. Thus, if dynamic simulations are to be used, then the model should be calibrated on dynamic data. The use of data from composite sample data from dynamic experiments will give a less accurate calibration, and the use of data from steady-state experiments will give the most risky calibration in this case. If, on the other hand, steady-state simulations are the objective, then steady-state data should be used for the calibration. The constants resulting from the calibration in the two cases mentioned above, need not – and will not in most cases – be identical.

Below, the two main levels of calibration are presented. They are based on the use of non-dynamic data (Level 1) and dynamic data (Level 2).

Level 1. Non-dynamic data. It is possible to make a calibration based on non-dynamic data, either from steady-state experiments or from (24 hours) composite samples from dynamic experiments. A compartmentalized flow scheme must be used, as such a system can give data with, and without, substrate limitation. This allows for calibration of some of the saturation coefficients. In the case of pilot-plant experiments, the calibration will be strongest if the pilot plant is compartmentalized in a similar manner to the full-scale plant. Pilot (or lab) experiments for calibration of a full-scale process, which is not intended to be compartmentalized, can still be run compartmentalized for short test periods, as this will not influence the biomass composition. The experiments in the pilot or the lab plant should thus be performed with the full-scale layout, but then for short test periods be modified to include compartmentalized reactors (Larrea *et al.*, 1992).

Wastewater COD (dichromate), TN and TP (all soluble and suspended) in the influent and in the effluent, plus NH_4^+ in the effluent from a compartmentalized system of ideal mix tanks are needed. In addition to this, oxygen uptake rate (OUR) measurements in each tank are required. Composite samples (or grab samples from a steady-state operation) in a process with nitrification, denitrification and biological phosphorus removal can be used. This information allows for calibration of the growth rate constants and saturation coefficients for heterotrophs, phosphotrophs and autotrophs.

Level 1a. As Level 1, but with sludge production figures or MLVSS (mixed liquor volatile suspended solids) concentrations measured. This allows for calibration of some stoichiometric constants and more detailed wastewater characterization.

Level 2. Dynamic experiments. Wastewater COD fractions, TN and TP (soluble and suspended) in the influent and in the effluent, and NH_4^+ and ortho-phosphate in the effluent are needed. Measurements of oxygen respiration rates are also required. All measurements should be taken from dynamic load (and hopefully dynamic response) experiments with nitrification, denitrification and biological phosphorus uptake. This information allows for calibration of the growth and hydrolysis constants for heterotrophs, phosphotrophs and autotrophs.

Level 2a. As Level 2 and with respiration and removal rate studies (oxygen uptake rate (OUR), ammonia uptake rate (AUR), nitrogen uptake rate (NUR) and phosphorus uptake rate (PUR)), in the biomass. This allows for a safer calibration of growth rates for autotrophs, heterotrophs, denitrifying biomass and phosphotrophs. Additional measurements of oxygen and nitrate respiration rates in the wastewater will also make it possible to characterize the fraction of heterotrophs and denitrifiers in the influent.

If BOD measurements are available, they can be used. But they need to be converted to COD before the simulation. BOD values used in simulations, without prior conversion to COD, will not respect the conservation principles on which stoichiometry is based. This will give erroneous results in the simulations.

5.2 Calibration using non-dynamic data (Level 1)

Calibration which is based on measurements of influent and effluent in a series of tanks, and which enables calibration of autotrophic, heterotrophic, denitrifying and phosphotrophic bacteria.

The soluble COD fractions (S_A and S_F), NH_4^+, NO_3^- and PO_4^{3-} in the effluent from well-planned experiments can be used to calibrate the growth kinetics.

Heterotrophic calibration. The calibration of the model parameters μ_H, b_H, $K_A(H)$ and $K_{O_2}(H)$ should be based on an analysis of what is the limiting factor in the experiments performed. If soluble degradable COD values and oxygen values are much higher (more than five times) than the normal range of the saturation coefficients, then the calibration can only be made via μ_H. A model which is not limited by the Monod terms in the reaction rate equations is in general difficult to calibrate, and the simulations tend to be less reliable. For paired parameters like μ_H and b_H, only the parameter with the highest relative influence (in this case μ_H) should be changed. Results from OUR measurements might be used to calibrate b_H (and μ_H).

Normally the Monod term for oxygen or COD will be the rate-limiting one. In this case the procedure for calibration is the following:

μ_H should normally not be changed. b_H should not be changed. If the S_A concentration is limiting the process, and the observed S_A is lower than the simulated one, then $K_A(H)$ should be decreased in order to increase the value of the Monod term for the simulation. The $K_A(H)$ interval will typically be 3–5 g COD m^{-3}, but values up to 20 g COD m^{-3} have been used, often coupled to high μ_H values (6–8 d^{-1}).

$K_A(H)$ is related to diffusion limitation in the flocs. In experiments with high turbulence and small flocs (as often found in pilot-scale experiments), the K_A-value tends to be low. In full-scale installations the K_A-value will often be higher, due to larger flocs and less turbulence.

$K_{O_2}(H)$ should not be changed, unless the experiments have been made in a DO-range where oxygen is limiting the process rate, i.e. 0–2 ppm. Considerations similar to those for $K_A(H)$ should be applied for the $K_{O_2}(H)$-value.

Autotrophic calibration. Nitrification can be calibrated using the procedure given below, which is similar to the procedure recommended for the heterotrophic growth constants. Ammonia in the effluent can be used to calibrate the autotrophic kinetics. The calibration of the model parameters μ_{AUT}, b_{AUT}, $K_{NH_4}(AUT)$ and $K_{O_2}(AUT)$ should be based on an analysis of what is the limiting factor in the experiments performed. If ammonia values and oxygen values are much higher (more than five times) than the normal range of the saturation coefficients, then the calibration can only be made via μ_{AUT} and b_{AUT}. In this case only μ_{AUT} should be changed.

If experiments have been made where the ammonia concentration has been limiting, then the following calibration procedure should be applied:

b_{AUT} should not be changed. μ_{AUT} should not be changed unless it is impossible to fit the effluent NH_4^+ values by changing the $K_{NH_4}(AUT)$. $K_{NH_4}(AUT)$ should only be changed if experiments in the range of 0–1.5 g NH_4^+-N m^{-3} have been performed, and after the growth rate has been changed. The value depends on actual turbulence and floc size distribution. $K_{O_2}(AUT)$ should only be changed if experiments in the range of 0–2 g O_2 m^{-3} have been performed, and after the growth rate has been changed. The value depends on turbulence and floc size.

Denitrification calibration. If denitrification is included in the experiments, then denitrification constants can be calibrated as follows:

In general μ_H should not be changed, especially if it has been calibrated by OUR measurements, or used to calibrate the COD values already. Changing the μ_H in this case would result in starting the total calibration procedure all over again. η_{NO_3} is calibrated based on the general level of nitrate in the effluent from the anoxic tank (typical values are 0.6–0.9). If the μ_H value has been fixed earlier, the calculated effluent nitrate concentration can be increased/decreased by decreasing/increasing the η_{NO_3} value. The calibration of denitrification will include the denitrification performed by the PAOs, a biomass fraction not modelled separately. The use of η_{NO_3}, instead of μ_H, for the calibration of denitrification is recommended when denitrifying PAOs are known to be a significant fraction of the PAOs. The saturation coefficient for denitrification, K_{NO_3}, should be treated with a procedure similar to that used for nitrification.

Calibration of biological phosphorus uptake. The calibration of the phosphotroph kinetics based on steady-state experiments seems to be rather risky at present, due to lack of experience with the variations of the parameters from place to place.

Y_{PO_4} can be calibrated by using observed phosphate concentrations in the effluent from the first anaerobic tank. If more than one anaerobic tank has been used, the following tanks can be used to calibrate fermentation. For tanks with high S_F, the maximum rate of fermentation, q_{fe}, can be changed until the simulated effluent phosphate concentration fits the observed data. For tanks with low S_F, K_{fe} can be calibrated. Aerobic growth parameters for the phosphotrophs, μ_{PAO} and K_P, can be calibrated. If OUR measurements are not available then μ_{PAO} should not be changed, unless changes in K_P cannot give a reasonable fit to the phosphate data.

5.3 Calibration using dynamic data (Level 2)

Dynamic experiments can be used to calibrate the phosphorus-accumulating bacteria kinetics (as well as the kinetics of the other biomass components). Procedures as decribed above should be used.

A detailed speciation of the influent and effluent COD will allow the determination of the following:

S_I, the inert soluble COD in the influent

S_A, the volatile acids/fermentation products

S_F, the easily (fermentable) degradable COD.

In combination with modelling, the following components can be determined:

X_S, the slowly degradable COD

X_H, the heterotrophic biomass

X_I, the inert suspended COD.

Calibration based on experiments with the biomass from the process allows for a very reliable calibration of the processes. The experiments can be batch or continuous. Respiration and uptake rates such as OUR, AUR, NUR and PUR can be used to calibrate maximum growth rates for heterotrophs, autotrophs, denitrifiers and phosphotrophs, and to calibrate the influent wastewater.

The calibration of phosphotrophs follows a procedure similar to the one described above:

Y_{PO_4} is calibrated by using the anaerobic influent and effluent ortho-phosphate and acetic acid concentrations. If the experiments are made with a series of anaerobic tanks, then the first tanks can be used to calibrate Y_{PO_4} and the last anaerobic tank to calibrate η_{fe} . q_{fe} and K_{fe} can only be calibrated as a pair and only in the case of experiments where the influent acetic acid has been used up in the first part of the anaerobic tank. This will make the fermentation-generated acetic acid the driving force for the phosphorus release in the last part of the anaerobic tank. Data from the aerobic tanks can be used to calibrate μ_{PAO} and K_P.

Experiments for calibration of the phosphotroph kinetics have to be very detailed and extremely well planned. Even for well-planned experiments they will be very time-consuming and expensive.

5.4 Calibration of temperature dependency

The various biological processes in **ASM2** have different temperature dependencies, which are normally described by exponential expressions such as $a^{(T-20)}$. Calibration experiments must be long term (at least three times the solid retention time), in order to ensure that the biomass composition is stable. The exponential expression will normally fit experimental data reasonably within an interval of 10 °C. For larger intervals the fit might be problematic, either at the low temperatures of the interval or at the high temperatures. Care should be taken when extrapolating outside the interval investigated. Extrapolations for more than 5 °C should not be made. The temperature dependencies for the processes in **ASM2** can be placed in four groups, as shown in Table 5.2.

Table 5.2. Temperature dependency, $a^{(T-20)}$, of processes in ASM2.

Degree of dependency	a	Processes
None	1.00	Chemical precipitation
Low	1.04	Phosphotrophs, hydrolysis
Medium	1.07	Heterotrophs, fermentation
High	1.12	Nitrification

Table 3.3 shows values of the kinetic constants at 10 and 20 °C. Many of the constants in the table, which are shown to have no temperature dependency, are more-or-less temperature dependent. The degree of their dependency is not well known, and for most model simulations this will have a minor impact only. In general the **ASM2** is valid for temperatures between 10 and 25 °C. Outside this interval the behaviour of the processes might be different. Thus care should be taken and temperature calibrations should be made.

6. Model limitations

The **Activated Sludge Model No. 2 (ASM2)** is developed based on its predecessor, the Activated Sludge Model No. 1 (**ASM1**). It is a prerequisite for the users of **ASM2** to have read the report on the previous model (Henze *et al.*, 1987), in which detailed explanations are given on basic and important aspects of the model, such as the method of model presentation (the matrix notation), the structure of the model incorporating carbon oxidation/nitrification/denitrification, the wastewater characterization, and the implementation of the activated sludge model.

It is essential for the users of the model to have a sufficient understanding of the model before using it. The Task Group members are responsible for the model structure itself, but its application and the interpretation of the simulation results are the users' responsibility. By changing the kinetic or stoichiometric parameters, by modifying the rate equations or the stoichiometry, or by adding/deleting something to/from the model, the users can adapt the model to the situations which they want to simulate. However, it is very dangerous to make such changes without a detailed understanding of the model; they may lead to erroneous results. The users should always try to understand why the model behaves in a particular way. If the users cannot follow the way in which the model has behaved, they should not rely on the simulation results.

The assumptions and restrictions associated with **ASM1**, which are described in the earlier report, are also applicable to the present model. Namely, constant values of the pH, the coefficients in the rate equations and the stoichiometry.

Some assumptions in **ASM1** have been extended in order to deal with biological phosphate removal, as follows:

- Heterotrophic biomass and phosphate-accumulating biomass are homogeneous and do not undergo changes with time, which is inherent in the assumption of constant kinetic parameters.
- Hydrolysis of organic matter, organic nitrogen and organic phosphate are coupled and occur simultaneously.

6.1 Assumptions regarding phosphate-accumulating organisms (PAOs)

In order to develop mathematical models for biological excess phosphate removal processes, it is essential to understand characteristics of the microorganisms responsible for the phosphate removal, the phosphate-accumulating organisms (PAOs), and their relationship with other bacterial populations. Despite extensive research efforts, the behaviour and physiology of PAOs in the biological phosphate removal processes have not yet been fully understood. In the present model, the heterotrophic biomass has been split into two fractions: heterotrophs and phosphate- accumulating organisms (PAOs). PAOs are defined as a group of bacteria that, in activated sludge processes with anaerobic and aerobic zones, exhibit biological phosphorus uptake. The following assumptions have been made with regard to the characterization of the biomass, including the PAOs.

1. Fermentation products such as acetate (S_A) are assumed to be the only organic substrates that can be taken up by the PAOs operating in the biological excess phosphorus removal mode. The (other) heterotrophic organisms are assumed to utilize fermentable organic substrates, S_F, as well as S_A. Short-chain fatty acids, which are classified into S_A in the present model, are known to be preferred carbon sources for biological phosphorus removal (Wentzel *et al.*, 1990). Therefore this assumption may be close to reality.
2. It is assumed that the PAOs can grow aerobically on stored PHA only, not on S_A directly. In spite of this assumption, the PAOs in the model may take up S_A under aerobic conditions without growth but with phosphate release. The produced PHA may then be utilized for growth. Thus, in terms of the model, there is competition for S_A under aerobic conditions between the PAOs and the (other) heterotrophs. The description of this competition still needs refinement. In the present model, this competition for S_A will be to the advantage of the heterotrophs, with their high growth rate. Thus, systems with significant input of S_A to the aerobic tanks should not be modelled with the present model, as erroneous results may be obtained.

3. It is assumed that the PAOs do not possess denitrifying capability. This assumption influences the modelling of the processes in the anaerobic tanks and the anoxic tanks.

   Anaerobic tanks. It is well known that the input of nitrate (NO_3^--N) to the anaerobic tank of biological phosphate removal systems can reduce the phosphate removal efficiency. There are three possible mechanisms for this type of phosphate removal deterioration.

   The first mechanism is a simple competition for S_A between the PAOs and the denitrifying fraction of the heterotrophs. Thus less S_A will be stored as PHA by the PAOs.

   The second mechanism is reduction of the PAO activity due to reduced fermentation in the anaerobic tank. Heterotrophs which denitrify do not ferment S_F to S_A. Thus less S_A is available in the anaerobic tank. This will reduce the amount of PHA accumulated in the anaerobic tank, which in turn will reduce the growth of the PAOs in the aerobic tank. Ultimately, this might lead to wash-out of the phosphotrophs.

   The third mechanism is the well-known fact that some of the PAOs can denitrify. With nitrate input to the 'anaerobic' tank, these will denitrify to get energy instead of utilizing poly-phosphate. The necessity to accumulate poly-phosphate disappears in this fraction of the PAOs and consequently the phosphate removal may deteriorate.

   All three mechanisms are realistic in actual situations, but only the first two are considered in the present model. The phosphorus removal deterioration, under conditions where nitrate exists, can be well modelled in most cases of domestic wastewater treatment, by assuming the two first mechanisms to be active only. The third mechanism, denitrifying PAOs, has not been included, due to lack of experience with these organisms and a wish to keep the model complexity at a minimum. Denitrifying PAOs will, in **ASM2**, be modelled as denitrifying heterotrophs. Small inputs of nitrate to the anaerobic tanks can be modelled without problems. Simulations of processes with high nitrate input to the anaerobic tanks may give erroneous results.

   Anoxic tanks. The assumption in the model that PAOs cannot denitrify means that an anoxic tank is anaerobic from a PAO point of view. PAOs will thus release phosphorus. In practice, release as well as uptake of phosphorus has been observed, which can be explained by a denitrifying fraction of the PAOs. These organisms have not been included in the present model. For modelling the anoxic tank, it might be necessary to increase the heterotrophic rate of denitrification by increasing η_{NO_3} (normally μ_H cannot be increased because of other calibration limitations).

4. Poly-hydroxy-alkanoates (PHAs) represent all the carbon storage materials in PAO cells in the present model, although glycogen or carbohydrate has been proposed as another carbon storage material, which provides reducing power for the synthesis of PHA under the anaerobic conditions in the biological phosphate removal processes (Mino *et al.*, 1987; Satoh *et al.*, 1992). Since the behaviour of intracellular carbohydrate is not sufficiently understood, the present model does not incorporate carbohydrate as an independent parameter. As for domestic wastewater treatment processes being operated under typical conditions, the carbohydrate content of the sludge is not a limiting factor for PHA synthesis.

5. The model does not include a separate biomass fraction of heterotrophs that can store PHA without phosphate release. The presence of such organisms can be included in a calibration by reducing the values of Y_{PO_4} and q_{PP}. The lack of this biomass fraction is not believed to be a significant limitation to the use of **ASM2**.

6. Heterotrophs defined in **ASM2** are assumed to grow aerobically, denitrify anoxically and ferment anaerobically. This kind of heterotrophic microorganism is not typical in a microbiological sense. A microbiologically sound classification of heterotrophs might be as follows:
   - Obligate aerobes
   - Aerobes with denitrifying capability
   - Facultative aerobes which cannot denitrify but ferment

The distribution of bacterial populations, like the above-mentioned bacterial groups, should affect the rate of different processes such as aerobic growth, denitrification and fermentation. In the present model the activities of different bacterial populations are reflected in the rate coefficients of the processes in which they are involved. By selecting appropriate rate coefficients, the variation in the bacterial populations from one location to another can be properly expressed, even though there is only one kind of heterotrophic organism.

6.2 Restrictions due to model structure

The following represent some of the restrictions associated with the structure of the model.

1. Low phosphate and ammonia concentrations. Since some inorganic nutrients like nitrogen, phosphorus and alkalinity (which in the model is treated as a 'nutrient') appear in the rate equations and the stoichiometry, they are assumed to be essential components for the processes in which they are involved. However, we do not know the detailed mechanisms of growth limitation caused by low nutrient concentrations in activated sludge systems. The model includes Monod terms that will limit the biomass growth at low nutrient concentrations. Care must be taken to be sure that sufficient

quantities of inorganic nutrients are present to allow for balanced growth. In some biological phosphorus processes, where S_{PO_4} is very low, phosphorus might be growth-limiting.

2. The fractions of organic substrates (S_A, S_F and X_S) are assumed to be homogeneous, and their nature should not change at all regardless of the type of compounds included in each fraction. Some particular organic compounds may have significant effects on phosphate removal. For example, it is reported that glucose-containing (non-domestic) wastewater causes phosphate removal deterioration due to significant growth of 'G bacterium', which under anaerobic conditions can take up organic substrates without utilizing poly-phosphate (Cech and Hartman, 1990). Since glucose is just a part of S_F and not a separate parameter, such a process cannot be simulated by the present model.
3. The effects of limitations of potassium and magnesium on biological phosphorus removal are not considered. It is well known that potassium and magnesium are the two major cations which make up poly-phosphate salts in PAOs. The shortage of these cations can lead to the deterioration of poly-phosphate accumulation in the sludge, thus to the drop in phosphorus removal.
4. Nitrite (NO_2^-) and nitrogen monoxide (NO) are reported to have inhibitory effects on biological excess phosphorus removal processes (Matsuo, 1988), but here such effects are not considered.

6.3 Constraints for useful simulations

The following represent some of the constraints which must not be violated if simulation results are to be usable in practice.

1. The present model is designed for domestic wastewater treatment only. It should not be applied to wastewaters containing significant industrial discharges. For example, the model does not consider carbohydrate-rich wastewaters or wastewaters containing some unusual compounds which are toxic or partially degradable.
2. **ASM2** does not deal with solids separation in settling tanks. It simulates processes in biological reactors, including the parts of the settling tanks that are biologically active.
3. The pH should be near neutrality, preferably within a range from 6.3 to 7.8. The charge balance calculation or the alkalinity balance calculation in the model is based on a pH value of 6.86. In other words, S_{ALK} is assumed to be bicarbonate, HCO_3^-, only, and S_{PO_4} is assumed to consist of 50% $H_2PO_4^-$ and 50% HPO_4^{2-}. In the practical application of the model, calculated low S_{ALK} values should be considered as a warning for possible low pH conditions.
4. The applicable temperature may have to be limited to a moderate range, probably from 10 to 25 °C. At higher or lower temperatures, the behaviour of PAOs is not fully understood and the model may not give reasonable predictions, especially for phosphate removal.

6.4 Important questions for further research

Of the processes included in **ASM2**, the following have the greatest need for further research. They are:

- Fermentation
- Anaerobic hydrolysis

Both processes are needed in a model dealing with biological phosphorus uptake. Few investigations into these processes have been made up to now, the reason being that it is difficult to create experiments where one of these two processes dominate, and which would allow for testing of kinetic expressions and calibrating kinetic constants. The two processes have been included in **ASM2** with kinetics which are as simple as possible.

Fermentation is believed to be the most important of the two in relation to biological phosphorus removal processes. Anaerobic hydrolysis is a slow process according to the default value for $\eta_{fe}(0.1)$, but this value has no substantive experimental basis.

With respect to parameter estimation and calibration, techniques need to be developed that will optimize such procedures for **ASM2**.

7. Conclusion

The **Activated Sludge Model No. 1 (ASM1)** has been shown to be a useful tool for research, development and optimization of biological nitrogen removal processes. In addition it has proved useful as a tool in teaching. The experience gained with **ASM1** has been used to develop a new model, **Activated Sludge Model No. 2 (ASM2)**, which incorporates biological phosphorus removal.

ASM2 provides a useful framework for further development of combined models for biological nitrogen and phosphorus removal.

As a research tool, it will help identify the processes and parameters of crucial importance in biological excess phosphorus removal, and focus attention on those aspects that require research and development.

Apart from research, the model can be used for process optimization, trouble-shooting and teaching.

As a design tool, however, the phosphorus part of the model has not yet achieved sufficient credibility for use.

ACTIVATED SLUDGE MODEL NO. 2d

by

IAWQ TASK GROUP ON MATHEMATICAL MODELLING FOR DESIGN AND OPERATION OF BIOLOGICAL WASTEWATER TREATMENT PROCESSES

Mogens Henze (Chairman), *Technical University of Denmark, Denmark*
Willi Gujer, *Swiss Federal Institute of Technology, Switzerland*
Takashi Mino, *Tokyo University, Japan*
Tomonori Matsuo, *Tokyo University, Japan*
Mark C. Wentzel, *Capetown University, South Africa*
Gerrit v. R. Marais, *Capetown University, South Africa*
Mark C.M. Van Loosdrecht, *Delft University of Technology, The Netherlands*

Contents

1. Introduction

This report presents a mathematical model which allows for dynamic simulation of combined biological processes for chemical oxygen demand (COD), nitrogen and phosphorus removal in activated sludge systems. The model as presented here is a tool for:

- Research (testing results, selecting and optimizing experiments)
- Process optimization and troubleshooting at full-scale treatment plants
- Teaching
- Design assistance (for optimization of details, not for full design)

The model presented below is not the final answer to biological phosphorus removal models. Rather it is a compromise between complexity and simplicity, and between the many viewpoints on what the correct model would be. It is intended to be a conceptual platform and reference for further model development.

ASM2d is an extension of the Activated Sludge Model No. 2 (Henze *et al.*, 1995) and the Activated Sludge Model No. 1 (ASM1) (Henze *et al.*, 1987), and uses the concepts incorporated in these models. ASM1 has since long proved to be an excellent tool for modelling nitrification-denitrification processes and has initiated further research in modelling and wastewater characterization. It is hoped that ASM2d will serve a similar function. ASM2d may be applied as presented, but based on experience, it will most likely be used as a platform for future model development. As this is the basic idea behind presenting the model, this is highly encouraged.

In ASM2 an unresolved part was the denitrification related to PAOs. Since the publication of ASM2 it has been demonstrated clearly (Mino *et al.*, 1995, Meinhold *et al.*, 1999, Kerrn-Jespersen and Henze, 1993) that PAOs in a modelling context can be considered to consist of two fractions, one of which can denitrify. This has created a need for an extension of ASM2, the result being presented here as ASM2d.

2. Conceptual approach

An attempt has been made to limit the number of processes used in the model. The aim has, however, been to produce a model that can reasonably describe the many different activated sludge system configurations which are used for biological phosphorus removal. This has resulted in the present level of complexity. In specific cases, it will be possible to reduce the complexity of the model by omitting processes that do not play a significant role, without interfering with the predictive power of the model.

The kinetics and stoichiometry used to describe the processes have been chosen as simply as possible, mainly based on Monod kinetics for all components that can influence the reaction rates. Monod kinetics allows for smooth transitions of the processes, as experience has shown. Kinetics and stoichiometry are presented using the matrix notation, which has been introduced together with ASM1 and appears at this moment to be the most efficient method to overview the complex transformations among the components. The matrix notation also allows control of the conservation of components in the stoichiometric coefficients and thus ensures that mass balances in the calculations are correctly maintained.

3. The Activated Sludge Model No. 2d

The Activated Sludge Model No. 2 (ASM2) is an extension of the Activated Sludge Model No. 1 (ASM1). ASM2 is more complex and includes many more components which are required in order to characterize the wastewater as well as the activated sludge. Additional biological processes are included, primarily in order to deal with biological phosphorus removal. The most significant change from ASM1 to ASM2 is the fact that the biomass now has cell internal structure, and therefore its concentration cannot simply be described with the distributed parameter X_{BM}. This is a prerequisite in order to include biological phosphorus removal in the model.

The Activated Sludge Model No. 2d is a minor extension of ASM2. It includes two additional processes to account for the fact that phosphorus accumulating organisms (PAOs) can use cell internal organic storage products for denitrification. Whereas ASM2 assumes PAOs to grow only under aerobic conditions, ASM2d includes denitrifying PAOs. This report is based on the previous report which introduced ASM2. All remarks made relative to ASM2 are equally valid for ASM2d. If information is given which relates specifically to ASM2d then reference will be made to this extended model.

In addition to the biological processes, ASM2 includes two 'chemical processes', which may be used to model chemical precipitation of phosphorus.

Whereas ASM1 was based entirely on COD for all particulate organic material, as well as the total concentration of the activated sludge, ASM2 includes poly-phosphates, a fraction of the activated sludge which is of prime importance for the performance of the activated sludge system, but which does not exert any COD. For this reason, the possibility of including total suspended solids (TSS) in the model is introduced. TSS also allow for inclusion of mineral particulate solids in the influent to treatment plants, as well as generation of such solids in the context of precipitation of phosphorus.

ASM2 is introduced here in a form which is more complex than a basic version, which could still predict many of the phenomena within a biological nutrient removal plant. The complex model as presented may easily be simplified by eliminating those components which do not have a dominant effect upon the kinetics of the processes, or the aspects of performance of the plant which are of interest.

ASM2 does not distinguish between the composition (cell internal structure) of individual cells but considers only the average composition of the biomass. Since each cell has a different history, its composition will typically deviate from the population average (e.g. it may not contain storage products whereas the average cell still has storage products available). This is of importance because kinetic expressions used in ASM2 are non-linear, and therefore average behaviour may not necessarily be predicted from average properties. In view of the additional problems that population models would introduce, the Task Group took the pragmatic decision to accept these problems and to propose ASM2 based on average properties of the population.

3.1 Components in the model

All symbols for model components distinguish between soluble '$S_?$' and particulate '$X_?$'. Within the activated sludge systems, particulate components, $X_?$, are assumed to be associated with the activated sludge (flocculated onto the activated sludge). They can be concentrated by sedimentation/thickening in clarifiers whereas soluble components, $S_?$, will only be transported with the water.

All particulate model components, $X_?$, must be electrically neutral (no ionic charges), soluble components, $S_?$, may carry ionic charge.

Soluble and particulate components may not necessarily be differentiated by filtration through 0.45 μm membrane filters as is frequently assumed in the technical literature. Some of these components are defined by their interaction with the biomass and require bioassays for their analysis (see Chapter 4 of the original report on ASM2 (Henze *et al.*, 1995))

All components are assumed to be homogeneous and distributed throughout the systems of interest.

3.1.1 Definition of soluble components, '$S_?$'

S_A **[M(COD) L^{-3}]: Fermentation products, considered to be acetate.** Since fermentation is included in the biological processes, the fermentation products must be modelled separately from other soluble organic materials. They are endproducts of fermentation. For all stoichiometric computations, it is assumed that S_A is equal to acetate, in reality a whole range of other fermentation products dominated by acetate is possible.

S_{ALK} **[mol(HCO_3^-) L^{-3}]: Alkalinity of the wastewater.** Alkalinity is used to approximate the conservation of electrical charges in biological reactions. Alkalinity is introduced in order to obtain an early indication of possible low pH conditions, which might inhibit some biological processes. For all stoichiometric computations, S_{ALK} is assumed to be bicarbonate, HCO_3^- only.

S_F **[M(COD) L^{-3}]: Fermentable, readily biodegradable organic substrates.** This fraction of the soluble COD is directly available for biodegradation by heterotrophic organisms. It is assumed that S_F may serve as a substrate for fermentation, therefore it does not include fermentation products.

S_I **[M(COD) L^{-3}]: Inert soluble organic material.** The prime characteristic of S_I is that these organics cannot be further degraded in the treatment plants dealt with in this report. This material is assumed to be part of the influent and it is also assumed to be produced in the context of hydrolysis of particulate substrates X_S.

S_{N_2} **[M(N) L^{-3}]: Dinitrogen, N_2.** S_{N_2} is assumed to be the only nitrogenous product of denitrification. S_{N_2} may be subject to gas exchange, parallel with oxygen, S_{O_2}.

S_{NH_4} **[M(N) L^{-3}]: Ammonium plus ammonia nitrogen.** For the balance of the electrical charges, S_{NH_4} is assumed to be all NH_4^+.

S_{NO_3} **[M(N) L^{-3}]: Nitrate plus nitrite nitrogen (NO_3^- + NO_2^--N).** S_{NO_3} is assumed to include nitrate as well as nitrite nitrogen, since nitrite is not included as a separate model component. For all stoichiometric computations (COD conservation), S_{NO_3} is considered to be NO_3^--N only.

S_{O_2} **[M(O_2) L^{-3}]: Dissolved oxygen.** Dissolved oxygen may be subject to gas exchange.

S_{PO_4} **[M(P) L^{-3}]: Inorganic soluble phosphorus, primarily ortho-phosphates.** For the balance of electrical charges, it is assumed that S_{PO_4} consists of 50% $H_2PO_4^-$ and 50% HPO_4^{2-}, independent of pH.

S_S **[M(COD) L^{-3}]: Readily biodegradable substrate.** This component was introduced in ASM1. In ASM2, it is replaced by the sum of $S_F + S_A$.

3.1.2 Definition of particulate components, '$X_?$'

X_{AUT} **[M(COD) L^{-3}]: Nitrifying organisms.** Nitrifying organisms are responsible for nitrification; they are obligate aerobic, chemo-litho-autotrophic. It is assumed that nitrifiers oxidize ammonium S_{NH_4} directly to nitrate S_{NO_3} (nitrifiers include both ammonium and nitrite oxidizers).

X_H **[M(COD) L^{-3}]: Heterotrophic organisms.** These organisms are assumed to be the 'all-rounder' heterotrophic organisms, they may grow aerobically and anoxically (denitrification) and be active anaerobically (fermentation). They are responsible for hydrolysis of particulate substrates X_S and can use all degradable organic substrates under all relevant environmental conditions.

X_I **[M(COD) L^{-3}]: Inert particulate organic material.** This material is not degraded within the systems of interest. It is flocculated onto the activated sludge. X_I may be a fraction of the influent or may be produced in the context of biomass decay.

X_{MeOH} **[M(TSS) L^{-3}]: Metal-hydroxides.** This component stands for the phosphorus-binding capacity of possible metal-hydroxides, which may be in the wastewater or may be added to the system. For all stoichiometric computations, it is assumed that this component is composed of $Fe(OH)_3$. It is possible to 'replace' this component with other reactants; this would require adaptation of the stoichiometric and kinetic information.

X_{MeP} **[M(TSS) L^{-3}]: Metal-phosphate, $MePO_4$.** This component results from binding phosphorus to the metal-hydroxides. For all stoichiometric computations, it is assumed that this component is composed of $FePO_4$. It is possible to 'replace' this component with other precipitation products; this would require adaptation of the stoichiometric and kinetic information.

X_{PAO} **[M(COD) L^{-3}]: Phosphate-accumulating organisms: PAO.** These organisms are assumed to be representative for all types of poly-phosphate-accumulating organism. The concentration of X_{PAO} does not include the cell internal storage products X_{PP} and X_{PHA}, but only the 'true' biomass. In ASM2d it is assumed that these organisms may grow in an anoxic as well as an aerobic environment whereas in ASM2 only aerobic growth is considered.

X_{PHA} [M(COD) L^{-3}]: A cell internal storage product of phosphorus-accumulating organisms, PAO. It includes primarily poly-hydroxy-alkanoates(PHA). It occurs only associated with X_{PAO}; it is, however, not included in the mass of X_{PAO}. X_{PHA} *cannot be directly compared* with analytically measured PHA concentrations; X_{PHA} is only a functional component required for modelling but not directly identifiable chemically. X_{PHA} may, however, be recovered in COD analysis, where it must satisfy COD conservation. For stoichiometric considerations, PHA is assumed to have the chemical composition of poly-β-hydroxy-butyrate $(C_4H_6O_2)_n$.

X_{PP} [M(P) L^{-3}]: Poly-phosphate. Poly-phosphate is a cell internal inorganic storage product of PAO. It occurs only associated with X_{PAO}; it is, however, not included in the mass of X_{PAO}. It is part of the particulate phosphorus and may be analytically observed. For stoichiometric considerations, poly-phosphates are assumed to have the composition of $(K_{0.33}Mg_{0.33}PO_3)_n$.

X_S [M(COD) L^{-3}]: Slowly biodegradable substrates. Slowly biodegradable substrates are high molecular weight, colloidal and particulate organic substrates which must undergo cell external hydrolysis before they are available for degradation. It is assumed that the products of hydrolysis (S_F) may be fermented.

X_{TSS} [M(TSS) L^{-3}]: Total suspended solids, TSS. Total suspended solids are introduced into the biokinetic models in order to compute their concentration via stoichiometry. Since phosphorus removal and precipitation introduce mineral fractions into the activated sludge, prediction of TSS is important.

3.2 Basis for the introduction of ASM2

3.2.1 Matrix notation

The Task Group introduced matrix notation for the presentation of biokinetic models in its report on the ASM1. The same concept will be used for the introduction of ASM2. It is assumed that the reader is familiar with this way of presenting biokinetics.

As a short summary: the components which are considered in the model and the transformation processes are characterized with the indices i and j respectively. Stoichiometric coefficients are presented in the form of a stoichiometric matrix $\nu_{j,i}$. The process rate equations form a vector ρ_j. The rate of production of the component i, r_i [M_i L^{-3} T^{-1}], in all parallel processes may then be computed from the sum:

$$r_i = \sum \nu_{j,i} \cdot \rho_j \quad \text{over all processes } j. \qquad (3.1)$$

Within the stoichiometric matrix one stoichiometric coefficient, $\nu_{j,k}$, of each process j may be chosen as dimensionless with the value of +1 or −1. For all other stoichiometric coefficients algebraic equations may be given, which introduce conservation principles into the determination of stoichiometric coefficients. Alternatively $\nu_{j,i}$ may be given in the form of absolute values with the dimension M_i M_k^{-1}, where M_k is the unit mass of the component k upon which stoichiometry is based (the component which has $\nu_{j,k} = +1$ or -1).

3.2.2 Conservation equations

Conservation equations are the mathematical equivalent of the principle that in chemical reactions, elements, electrons (or COD) and net electrical charges may neither be formed nor destroyed.

The stoichiometry of ASM1 is implicitly based on three conservation considerations for COD, electrical charges and nitrogen. ASM2 adds phosphorus conservation to these three. Further, an equation is introduced which converts the different solid components $X_?$ from their unit of measurement, to total suspended solids, X_{TSS}.

A conservation equation, which is valid for all processes j and all materials c subject to conservation, may be written as:

$$\sum \nu_{j,i} \cdot i_{c,i} = 0 \quad \text{over all components } i, \qquad (3.2)$$

where

$\nu_{j,i}$ = stoichiometric coefficient for component i in process j [M_i M_k^{-1}],

$i_{c,i}$ = conversion factor to convert the units of component i to the units of the material c, to which conservation is to be applied [M_c M_i^{-1}].

Each conservation equation contains *a priori* information and may be applied to each process. Each conservation equation allows the prediction of one stoichiometric coefficient without performing an experiment, provided the other coefficients are known.

In ASM2, these equations are used to

Table 3.1. Conversion factors $i_{c,i}$ to be applied in the conservation equation of ASM2. Missing values are equal to 0. The units of $i_{c,i}$ are $M_c\ M_i^{-1}$, e.g. $i_{N,2} = i_{NSF}$ g N g^{-1} COD or $i_{Charge,3} = -1/64$ moles$^+$ g^{-1} COD.

Index c: Factor index i:	Conservation for Component	Units	COD $i_{COD,i}$ g COD	N $i_{N,i}$ g N	P $i_{P,i}$ g P	Charge $i_{Charge,i}$ mole$^+$	Mass $i_{TSS,i}$ g TSS
1	S_{O_2}	g O_2	-1				
2	S_F	g COD	1	i_{N,S_F}	i_{P,S_F}		
3	S_A	g COD	1			-1/64	
4	S_{NH_4}	g N		1		+1/14	
5	S_{NO_3}	g N	-64/14	1		-1/14	
6	S_{PO_4}	g P			1	-1.5/31	
7	S_I	g COD	1	i_{N,S_I}	i_{P,S_I}		
8	S_{ALK}	mole HCO_3^-				-1	
9	S_{N_2}	g N	-24/14	1			
10	X_I	g COD	1	i_{N,X_I}	i_{P,X_I}		i_{TSS,X_I}
11	X_S	g COD	1	i_{N,X_S}	i_{P,X_S}		i_{TSS,X_S}
12	X_H	g COD	1	$i_{N,BM}$	$i_{P,BM}$		$i_{TSS,BM}$
13	X_{PAO}	g COD	1	$i_{N,BM}$	$i_{P,BM}$		$i_{TSS,BM}$
14	X_{PP}	g P			1	-1/31[a)]	3.23
15	X_{PHA}	g COD	1				0.60
16	X_{AUT}	g COD	1	$i_{N,BM}$	$i_{P,BM}$		$i_{TSS,BM}$
17	X_{TSS}	g TSS					-1[b)]
18	X_{MeOH}	g TSS					1
19	X_{MeP}	g TSS			0.205		1

All absolute numbers are obtained based on the chemical composition of the component (see definition of component). All factors i_{ci} are model parameters and must be obtained from experiments (See also Table 9).
a) Since ASM2 does not account for K^+ and Mg^{2+} this factor must compensate for their charge.
b) Since TSS are counted twice, this factor must be negative.

estimate the stoichiometric coefficients of S_{O_2} (S_{NO_3} and S_{N_2} in denitrification) from COD, S_{NH_4} from nitrogen, S_{PO_4} from phosphorus, S_{ALK} from charge and X_{TSS} from total solids conservation. Table 3.1 is a summary of the conversion factors $i_{c,i}$ which must be applied in Equation 3.2. These conversion factors are, wherever possible, obtained from chemical stoichiometry. 'COD' as a conservative property is defined as closely as possible to the analytically obtained COD. Examples are:

$i_{COD,5} = -64$ g O_2/14 g NO_3^--N from:
$NO_3^- + H_2O + 2\,H^+ \rightarrow NH_4^+ + 2\,O_2$

Or, one mole of nitrate (14 g N) has a negative oxygen demand ('liberates oxygen') of two moles of oxygen (64 g O_2). Similar arguments lead to:

$i_{COD,9} = -24$ g O_2/14 g N_2 from:
$2\,N_2 + 6\,H_2O + 4\,H^+ \rightarrow 4\,NH_4^+ + 3\,O_2$

All conversion factors given with absolute numbers in Table 3.1 may be obtained from chemical stoichiometry, based on the definition of the compounds. All factors identified with a symbol $i_{c,i}$ must be obtained from chemical analysis. Since ASM2 does not account for potassium (K^+) and magnesium ions (Mg^{2+}) X_{PP} must include these counterions. This is taken care of by the conversion factor $i_{Charge,14} = -1/31$.

As an example, the stoichiometric coefficient for component 2 (i = 2) in the third process (j = 3) may be obtained from the conservation equation for COD based on Equation 3.2 according to:

$$\nu_{3,2} = -(\nu_{3,1} \cdot i_{COD,1} + \nu_{3,3} \cdot i_{COD,3} + \ldots + \nu_{3,n} \cdot i_{COD,n}) / i_{COD,2}$$

or

$$\nu_{3,2} = -\left[\sum_i (\nu_{3,i} \cdot i_{COD,i}) - \nu_{3,2} \cdot i_{COD,2}\right] \Big/ i_{COD,2}.$$

The introduction of the conservation equations in an abstract form may at first appear to be complicated. However, the concept is directed towards its application in computer programs and helps to simplify the development of program code.

3.3 Biological processes, stoichiometry and kinetics

The biological processes of ASM2 are introduced here. A full stoichiometric matrix using typical stoichiometric coefficients is presented in Table 4.4.

Table 3.2. Stoichiometry of hydrolysis processes. The stoichiometric parameters are defined in Table 4.2.

	Process	S_F	S_{NH_4}	S_{PO_4}	S_I	S_{ALK}	X_S	X_{TSS}
1	Aerobic hydrolysis	$1-f_{S_I}$	ν_{1,NH_4}	ν_{1,PO_4}	f_{S_I}	$\nu_{1,ALK}$	-1	$\nu_{1,TSS}$
2	Anoxic hydrolysis	$1-f_{S_I}$	ν_{2,NH_4}	ν_{2,PO_4}	f_{S_I}	$\nu_{2,ALK}$	-1	$\nu_{2,TSS}$
3	Anaerobic hydrolysis	$1-f_{S_I}$	ν_{3,NH_4}	ν_{3,PO_4}	f_{S_I}	$\nu_{3,ALK}$	-1	$\nu_{3,TSS}$

The stoichiometric coefficients for S_{NH_4}, S_{PO_4}, S_{ALK} and X_{TSS} may be computed from Conservation Equation 3.2 with the aid of Table 3.1. As an example $\nu_{1,PO_4} = -[(1-f_{S_I}) \cdot i_{PS_F} + f_{S_I} \cdot i_{PS_I} - 1 \cdot i_{PX_S}]/1$.

3.3.1 Biological processes, general remarks

Microorganisms have a complex cell internal structure and respond to different environmental conditions with adjustment of this structure. A frequently observed phenomenon is unbalanced growth, a situation where not all fractions of the cells are reproduced at an equal rate. Modelling such shifts of cell internal structure would require modelling of the different fractions of the biomass, a task which would be most fruitful if the behaviour of axenic cultures were described. Here, only three groups of microorganisms represent a vast variety of unknown species; each biological process described in ASM2 represents a large number of processes which act upon a variety of substances, which in the model are summarized in terms of COD.

Process descriptions in ASM2 are therefore based on the average behaviour of these different microorganisms, and are described in the way balanced growth processes would be modelled.

3.3.2 Hydrolysis processes

Many high molecular weight, colloidal or particulate organic substrates cannot be utilized directly by microorganisms. These substrates must be made available by cell external enzymatic reactions which are called hydrolysis processes. It is unclear whether the products of hydrolysis ever exist in true solution or whether they are taken up directly by the organisms which catalyse hydrolysis. Typically hydrolysis processes are considered to be surface reactions, which occur in close contact between the organisms which provide the hydrolytic enzymes and the slowly biodegradable substrates themselves.

Parallel with hydrolysis the activity of protozoa contribute to phenomena which are assigned to hydrolysis. Whereas it is difficult to distinguish between true hydrolysis and protozoan activity it is becoming more and more evident that the effect of electron acceptor upon the 'hydrolysis' process may actually be due to the inactivity of protozoa under anoxic and anaerobic conditions. Experimental evidence that 'hydrolysis' reactions depend on the available electron acceptors, leeds to the differentiation of three hydrolysis processes in ASM2. It is, however, a difficult task to estimate hydrolysis rate constants under different electron acceptor conditions.

1. Aerobic hydrolysis of slowly biodegradable substrate characterizes hydrolysis under aerobic conditions ($S_{O_2} > 0$).
2. Anoxic hydrolysis of slowly biodegradable substrate characterizes hydrolysis under anoxic conditions ($S_{O_2} \approx 0$, $S_{NO_3} > 0$). This process is typically slower than aerobic hydrolysis.
3. Anaerobic hydrolysis of slowly biodegradable substrate characterizes hydrolysis under anaerobic conditions ($S_{O_2} \approx 0$, $S_{NO_3} \approx 0$). This process is not well characterized and is probably slower than aerobic hydrolysis. Its rate remains to be studied.

Table 3.2 summarizes the stoichiometry of the hydrolysis processes. It is assumed that slowly biodegradable substrate X_S is degraded to readily degradable substrate S_F whereby a small fraction f_{S_I} of inert organic material S_I is released. The stoichiometric coefficients for S_{NH_4}, S_{PO_4} and S_{ALK} may be computed from Conservation Equation 3.2. These three coefficients are typically positive.

The proposed rate equations for the hydrolysis processes 1–3 are presented in Table 3.7. They are similar to those of ASM1: hyperbolic switching functions for S_{O_2} and S_{NO_3} consider the environmental conditions; a surface-limited reaction $(X_S/X_H)/(K_X + X_S/X_H)$ is assumed for the hydrolysis process itself. It is proposed that only heterotrophic organisms may catalyse hydrolysis. Typically hydrolysis is slower under denitrifying or anaerobic (fermentation) than under aerobic conditions. The rate for anoxic and anaerobic hydrolysis is therefore reduced by the factors η_{NO_3} and η_{fe} respectively.

The hydrolysis of particulate, biodegradable organic nitrogen is included as a separate process in ASM1 but not in ASM2. This

Table 3.3. Stoichiometry of the facultative heterotrophic organisms X_H. The stoichiometric parameters are defined in Table 4.2. Stoichiometry for S_{O_2}, S_{NH_4}, S_{PO_4}, S_{ALK} and X_{TSS} may be computed from conservation.

	Process	S_{O_2}	S_F	S_A	S_{NO_3}	S_{N_2}	X_I	X_S	X_H
4	Aerobic growth on S_F	$1-\frac{1}{Y_H}$	$-\frac{1}{Y_H}$						1
5	Aerobic growth on S_A	$1-\frac{1}{Y_H}$		$-\frac{1}{Y_H}$					1
6	Anoxic growth on S_A		$-\frac{1}{Y_H}$		$-\frac{1-Y_H}{2.86\cdot Y_H}$	$\frac{1-Y_H}{2.86\cdot Y_H}$			1
7	Anoxic growth on S_A, Denitrification			$-\frac{1}{Y_H}$	$-\frac{1-Y_H}{2.86\cdot Y_H}$	$-\frac{1-Y_H}{2.86\cdot Y_H}$			1
8	Fermentation		-1	1					
9	Lysis						f_{X_I}	$1-f_{X_I}$	-1

process is necessary if the nitrogen content of X_S is variable. In order to simplify ASM2, it is assumed that X_S contains a constant fraction of nitrogen i_{N,X_S} and phosphorus i_{P,X_S}. Without this simplifying assumption, six more hydrolysis processes and two more particulate components would be required.

The process of ammonification is included in ASM1 in order to describe the release of ammonium, S_{NH_4}, from soluble, biodegradable organic nitrogen. In ASM2 it is assumed that the fermentable substrates, S_F, contain a constant fraction of nitrogen and phosphorus, i_{N,S_F} and i_{P,S_F} respectively. This allows the process of ammonification to be ignored. Without this simplifying assumption, two more processes (ammonification as well as phosphatification, the release of phosphate S_{PO_4} from an organic fraction), and two more components (soluble, degradable organic nitrogen and phosphorus) would have to be introduced.

3.3.3 Processes of facultative heterotrophic organisms

The heterotrophic organisms X_H are responsible for the hydrolysis of slowly biodegradable substrate X_S (see above), the aerobic degradation of fermentable organic substrates S_F and of fermentation products S_A (aerobic growth), anoxic oxidation of S_F and S_A and reduction of nitrate S_{NO_3} (denitrification), and anaerobic fermentation of S_F to S_A. In addition these organisms are subject to decay and lysis. The stoichiometry and the kinetics of the processes described below are presented in Tables 3.3 and 3.7 respectively.

4 and 5. Aerobic growth of heterotrophic organisms on fermentable substrates S_F and on fermentation products S_A. These processes are modelled as two parallel processes, which consume the two degradable organic substrates S_F and S_A. For both processes identical growth rates μ_m and yield coefficients Y_H are assumed. The rate equations are designed such that the maximum specific growth rate of the heterotrophic organisms does not increase above μ_m even if both substrates, S_F and S_A, are present in high concentrations. These processes require oxygen, S_{O_2}, nutrients, S_{NH_4} and S_{PO_4}, and possibly alkalinity, S_{ALK}, and they produce suspended solids, X_{TSS}.

6 and 7. Anoxic growth of heterotrophic organisms on fermentable substrates, S_F, and on fermentation products, S_A; denitrification. These two processes are similar to the aerobic growth processes, but they require nitrate, S_{NO_3}, as the electron acceptor rather than oxygen. The stoichiometry for nitrate is computed based on the assumption that all nitrate, S_{NO_3}, is reduced to dinitrogen, S_{N_2}. Denitrification releases alkalinity, the stoichiometry of which is predicted from charge conservation. Denitrification is assumed to be inhibited by oxygen S_{O_2} and the maximum growth rate μ_m is reduced relative to its value under aerobic conditions, by the factor η_{NO_3}. This accounts for the fact that not all heterotrophic organisms X_H may be capable of denitrification or that denitrification may only proceed at a reduced rate.

8. Fermentation. Under anaerobic conditions ($S_{O_2} \approx 0$, $S_{NO_3} \approx 0$) it is assumed that heterotrophic organisms are capable of fermentation, whereby readily biodegradable substrates S_F are transformed into fermentation products S_A. Although this process may possibly cause growth of heterotrophic organisms, it

Table 3.4. Stoichiometry of the phosphorus-accumulating organisms, PAO, for ASM2d. The stoichiometric parameters are defined in Table 4.2. Stoichiometry for S_{O_2}, S_{NH_4}, S_{N_2}, S_{NO_3}, S_{PO_4}, S_{ALK} and X_{TSS} may be computed from conservation. ASM2 does not include processes 12 and 14.

	Process	S_{O_2}	S_A	S_{N_2}	S_{NO_3}	S_{PO_4}	X_I	X_S	X_{PAO}	X_{PP}	X_{PHA}
10	Storage of X_{PHA}		-1			Y_{PO_4}				$-Y_{PO_4}$	1
11	Aerobic storage of X_{PP}	$-Y_{PHA}$				-1				1	$-Y_{PHA}$
12	Anoxic storage of X_{PP}			$-\nu_{12,NO_3}$	ν_{12,NO_3}	-1				1	$-Y_{PHA}$
13	Aerobic growth of X_{PAO}	ν_{13,O_2}				$-i_{PBM}$			1		$-1/Y_H$
14	Anoxic growth of X_{PAO}			$-\nu_{14,NO_3}$	ν_{14,NO_3}	$-i_{PBM}$			1		$-1/Y_H$
15	Lysis of X_{PAO}					ν_{15,PO_4}	f_{X_I}	$1-f_{X_I}$	-1		
16	Lysis of X_{PP}					1				-1	
17	Lysis of X_{PHA}		1								-1

is introduced here as a simple transformation process. A growth process would require more complex kinetics, more kinetic and stoichiometric parameters which are difficult to obtain, and possibly different yield coefficients for S_F and S_A in processes 4 to 7. Fermentation releases negatively charged fermentation products, S_A, and therefore has a requirement for alkalinity, S_{ALK}. This is predicted from charge conservation.

Fermentation is a process which, up to now, has not been well characterized. Little is known about the kinetics of this process, which may lead to a large range of kinetic parameters for modelling experimental results. Reliable application of ASM2 requires that research is directed towards characterizing what is described here with the process of fermentation.

9. Lysis of heterotrophic organisms. This process represents the sum of all decay and loss processes of the heterotrophic organisms: endogenous respiration, lysis, predation etc. It is modelled in analogy to ASM1; its rate is independent of environmental conditions.

3.3.4 Processes of phosphorus-accumulating organisms

Some organisms, X_{PAO}, are known for their potential to accumulate phosphorus in the form of poly-phosphate X_{PP}. Currently these organisms are not well characterized; historically it was assumed that they would all be part of the *Acinetobacter* genus. However, today it is clear that *Acinetobacter* may contribute to, but do by far not dominate, biological phosphorus removal. Initially it was assumed that phosphorus-accumulating organisms, PAO, could not denitrify; now evidence has become available that some of them can denitrify. Phosphate release is sometimes slower in the presence of nitrate; this observation is not predicted with ASM2 but is included in ASM2d. Glycogen is found to be an important carbon storage material of PAO but is not considered in ASM2 in order to reduce model complexity. This restriction leads to limitations of the applicability of ASM2d which will be discussed later.

The greater the attempts to characterize PAO, the more complex this group of organisms becomes. The Task Group is well aware that the time has come when biological phosphorus removal is being designed and used in actual plants. The introduction of a very detailed mechanistic model for the processes responsible for biological phosphorus removal is, however, premature. The Task Group therefore has chosen to suggest a simple model, which allows prediction of biological phosphorus removal, but does not yet include all observed phenomena. The model proposed may be the base for further development. With the introduction of ASM2d the most important criticism that PAO contribute significantly to denitrification which is not described in ASM2 is taken care of.

The following model for the behaviour of phosphorus-accumulating organisms, X_{PAO}, is valid for ASM2d only, it assumes that these organisms can grow under aerobic ($S_{O_2} > 0$) as well as anoxic ($S_{O_2} \approx 0$, $S_{NO_3} > 0$) conditions. They can only grow on cell internal stored organic materials, X_{PHA}. This assumption is a severe restriction of ASM2d and may lead to further extensions. The stoichiometry and the kinetics of the processes described below are presented in Tables 3.4 and 3.7 respectively.

10. Storage of X_{PHA}. It is assumed that PAO may

release phosphate, S_{PO_4} from poly-phosphate, X_{PP}, and utilize the energy which becomes available from the hydrolysis of X_{PP}, in order to store cell external fermentation products S_A in the form of cell internal organic storage material X_{PHA}. The process is primarily observed under anaerobic conditions. However, since the process has also been reported to occur under aerobic and anoxic conditions, the kinetic expression does not include inhibition terms for S_{O_2} and S_{NO_3}. Experimental observation of this process is easy if the release of phosphorus is observed rather than the organics which are stored. Experience indicates, however, that the rate of storage of organics is relatively constant, whereas the release of phosphorus varies, indicating a variable stoichiometric relationship. The base for the stoichiometry of this process was therefore chosen to be the organics which are taken up, S_A and X_{PHA}. Reliable estimation of the rate constant, q_{PHA}, and the stoichiometric parameter, Y_{PO_4}, requires independent measurement of both S_A removal and S_P release. It has been shown that Y_{PO_4} depends on pH.

11 and 12. Aerobic and anoxic storage of poly-phosphate. Storage of ortho-phosphate, S_{PO_4}, in the form of cell internal poly-phosphates, X_{PP}, requires the PAO to obtain energy, which may be gained from the aerobic or anoxic respiration of X_{PHA}. The regeneration of poly-phosphates is a requirement for the growth of PAO, because the organic substrates, S_A, are stored only upon the release of poly-phosphate. Storage of X_{PP} is observed to stop if the phosphorus content of the PAO becomes too high. This observation leads to an inhibition term of X_{PP} storage, which becomes active as the ratio X_{PP}/X_{PAO} approaches the maximum allowable value of K_{MAX}. Under anoxic conditions the maximum rate of storage of poly-phosphate q_{PP} is reduced relative to its value under aerobic conditions, by the factor η_{NO_3}. This accounts for the fact that not all PAO (X_{PAO}) may be capable of denitrification or that denitrification may only proceed at a reduced rate. Process 12 is contained in ASM2d but not in ASM2.

13 and 14. Aerobic and anoxic growth of phosphorus-accumulating organisms. These organisms are assumed to grow only at the expense of cell internal organic storage products X_{PHA}. As phosphorus is continuously released by the lysis of X_{PP}, it is possible to assume that the organisms consume ortho-phosphate, S_{PO_4}, as a nutrient for the production of biomass. It is known that PAO may grow at the expense of soluble substrates (e.g. S_A), but it is unlikely that such substrates ever become available under aerobic or anoxic conditions in a biological nutrient removal plant. The Task Group therefore suggests this possibility to be ignored at this time. Under anoxic conditions the maximum growth rate of PAO μ_{PAO} is reduced relative to its value under aerobic conditions, by the factor η_{NO_3}. This accounts for the fact that not all PAO (X_{PAO}) may be capable of denitrification or that denitrification may only proceed at a reduced rate. Process 13 is contained in ASM2d but not in ASM2.

15, 16 and 17. Lysis of phosphorus-accumulating organisms and their storage products. Death, endogenous respiration and maintenance all result in a loss or decay of all fractions of PAO. Since the storage products X_{PP} and X_{PHA} are accounted for separately from the biomass X_{PAO}, all three components must be subject to separate decay processes. ASM2 includes three lysis processes which are all first-order relative to the component which is lost. If all three rate constants are equal, the composition of the organisms does not change due to decay. There is experimental evidence that X_{PP} decays faster than X_{PAO} and X_{PHA}. This additional loss of poly-phosphates may be modelled by the choice of an increased rate, b_{PP}, for the lysis of this component. The products of lysis are chosen in analogy to the lysis of heterotrophic organisms; storage products are assumed to decay to ortho-phosphate S_{PO_4} and fermentation products S_A.

3.3.5 Nitrification processes

Nitrification is assumed to be a one-step process, from ammonium S_{NH_4} directly to nitrate S_{NO_3}. The intermediate component, nitrite, is not included as a model component. In the context of nitrification, modelling nitrite production and consumption would be relatively easy. However, nitrite is also produced and consumed in the context of denitrification where the Task Group felt that the required addition to the model complexity does not warrant its inclusion at the present time. Modelling nitrite in nitrification but not in denitrification would, however, not be consistent and could lead to erroneous model predictions.

The stoichiometry and the kinetics of the processes described below, are presented in Tables 3.5 and 3.7 respectively.

Table 3.5. Stoichiometry of the growth and decay processes of nitrifying organisms X_{AUT}. The stoichiometric parameters are defined in Table 4.2. Stoichiometry for S_{O_2}, S_{NH_4}, S_{PO_4}, S_{ALK} and X_{TSS} may be computed from conservation.

	Process	S_{O_2}	S_{NH_4}	S_{NO_3}	S_{PO_4}	X_I	X_S	X_{AUT}
18	Aerobic growth of X_{AUT}	$-\frac{4.57 - Y_A}{Y_A}$	ν_{18,NH_4}	$\frac{1}{Y_a}$	$-i_{P,BM}$			1
19	Lysis		ν_{19,NH_4}		ν_{19,PO_4}	f_{X_I}	$1 - f_{X_I}$	-1

Table 3.6. Stoichiometry of the processes describing simultaneous precipitation of phosphorus. The absolute values of stoichiometry (and kinetics in Table 4.3) are based on the assumption that $Fe(OH)_3$ is used to precipitate S_{PO_4} in the form of $FePO_4$ + $Fe(OH)_3$. Stoichiometry for S_{ALK} and X_{TSS} may be computed from conservation.

	Process	S_{PO_4}	S_{ALK}	X_{MeOH}	X_{MeP}	X_{TSS}
20	Precipitation	−1	$\nu_{20,ALK}$	−3.45	4.87	1.42
21	Redissolution	1	$\nu_{21,ALK}$	3.45	−4.87	−1.42

18. Growth of nitrifying organisms. Nitrifying organisms are obligate aerobic, they consume ammonium as a substrate and a nutrient, and produce nitrate. Nitrification reduces alkalinity. The process is modelled as proposed in ASM1 with the exception of a phosphorus uptake into the biomass.
19. Lysis of nitrifying organisms. The process of lysis of nitrifiers is modelled in analogy to ASM1 and to the process of lysis of heterotrophic organisms. Since the decay products of lysis (X_S and ultimately S_F) are available substrates for heterotrophic organisms only, endogenous respiration of nitrifiers becomes manifest as an increased growth and oxygen consumption of heterotrophs. This is in analogy to ASM1.

3.4 Chemical precipitation of phosphates

In biological nutrient removal systems, metals, which are naturally present in the wastewater (e.g. Ca^{2+}), together with the high concentration of released soluble ortho-phosphate, S_{PO_4}, may result in chemical precipitation of phosphorus (e.g. in the form of apatite or calcium phosphate).

Further, simultaneous precipitation of phosphorus via the addition of iron or aluminium salts is a very common process for phosphorus removal worldwide. Simultaneous precipitation may be used in combination with biological phosphorus removal if the carbon to phosphorus ratio is unfavourably small.

In order to model the low effluent concentrations of ortho-phosphate, S_{PO_4}, which are observed in practice and which are partly due to chemical precipitation, the Task Group suggests a very simple precipitation model, which may be calibrated for a variety of situations. For this purpose, two processes (precipitation and redissolution) and two more components (X_{MeOH} and X_{MeP}) are added to ASM2. If chemical precipitation is not of any interest, these additions may be deleted from the model.

20 and 21. Precipitation and redissolution of phosphate S_{PO_4}. The precipitation model is based on the assumption that precipitation and redissolution are reverse processes, which at steady state would be in equilibrium according to:

$$X_{MeOH} + S_{PO_4} \leftrightarrow X_{MeP}$$

Precipitation and redissolution may be modelled with the following process rates respectively:

$$\rho_{20} = k_{PRE} \cdot S_{PO_4} \cdot X_{MeOH}$$
$$\rho_{21} = k_{RED} \cdot X_{MeP}$$

If both processes are in equilibrium ($\nu_{20,i} \cdot \rho_{20} = \nu_{21,i} \cdot \rho_{21}$) then an equilibrium constant may be derived as:

$$K_{eq} = \frac{\nu_{21,i} \cdot k_{RED}}{\nu_{20,i} \cdot k_{PRE}} = \frac{S_{PO_4} \cdot X_{MeOH}}{X_{MeP}}$$

Processes 20 and 21 are introduced here based on the assumption that X_{MeOH} and X_{MeP} are composed of ferric-hydroxide, $Fe(OH)_3$, and ferric-phosphate, $FePO_4$, respectively. This leads to the stoichiometry indicated in Table 3.6. The indicated rates of the processes result in residual ortho-phosphate concentrations, S_P, which at steady state are typical for simultaneous precipitation with the addition of $FeCl_3$. In this case, the addition of Fe^{3+} to the influent of a treatment plant may be modelled by the choice of X_{MeOH} in the influent recognizing that 1 g Fe^{3+} m^{-3} leads to 1.91 g $Fe(OH)_3$ m^{-3} = 1.91 g MeOH m^{-3} (which also increases influent X_{TSS} and decreases influent alkalinity S_{ALK}).

Table 3.7. Process rate equations for ASM2d. The kinetic parameters are defined in Table 4.3.

j	Process	Process rate equation ρ_j, $\rho_j \geq 0$ $[M_I\, L^{-3}\, T^{-1}]$
Hydrolysis processes:		
1	Aerobic hydrolysis	$K_h \cdot \frac{S_{O_2}}{K_{O_2}+S_{O_2}} \cdot \frac{X_S/X_H}{K_X+X_S/X_H} \cdot X_H$
2	Anoxic hydrolysis	$K_h \cdot \eta_{NO_3} \cdot \frac{K_{O_2}}{K_{O_2}+S_{O_2}} \cdot \frac{S_{NO_3}}{K_{NO_3}+S_{NO_3}} \cdot \frac{X_S/X_H}{K_X+X_S/X_H} \cdot X_H$
3	Anaerobic hydrolysis	$K_h \cdot \eta_{fe} \cdot \frac{K_{O_2}}{K_{O_2}+S_{O_2}} \cdot \frac{K_{NO_3}}{K_{NO_3}+S_{NO_3}} \cdot \frac{X_S/X_H}{K_X+X_S/X_H} \cdot X_H$
Heterotrophic organisms: X_H		
4	Growth on fermentable substrates, S_F	$\mu_H \cdot \frac{S_{O_2}}{K_{O_2}+S_{O_2}} \cdot \frac{S_F}{K_F+S_F} \cdot \frac{S_F}{S_F+S_A} \cdot \frac{S_{NH_4}}{K_{NH_4}+S_{NH_4}} \cdot \frac{S_{PO_4}}{K_P+S_{PO_4}} \cdot \frac{S_{ALK}}{K_{ALK}+S_{ALK}} \cdot X_H$
5	Growth on fermentation products, S_A	$\mu_H \cdot \frac{S_{O_2}}{K_{O_2}+S_{O_2}} \cdot \frac{S_A}{K_A+S_A} \cdot \frac{S_A}{S_F+S_A} \cdot \frac{S_{NH_4}}{K_{NH_4}+S_{NH_4}} \cdot \frac{S_{PO_4}}{K_P+S_{PO_4}} \cdot \frac{S_{ALK}}{K_{ALK}+S_{ALK}} \cdot X_H$
6	Denitrification with fermentable substrates, S_F	$\mu_H \cdot \eta_{NO_3} \cdot \frac{K_{O_2}}{K_{O_2}+S_{O_2}} \cdot \frac{K_{NO_3}}{K_{NO_3}+S_{NO_3}} \cdot \frac{S_F}{K_F+S_F} \cdot \frac{S_F}{S_F+S_A} \cdot \frac{S_{NH_4}}{K_{NH_4}+S_{NH_4}} \cdot \frac{S_{PO_4}}{K_P+S_{PO_4}} \cdot \frac{S_{ALK}}{K_{ALK}+S_{ALK}} \cdot X_H$
7	Denitrification with fermentation products, S_A	$\mu_H \cdot \eta_{NO_3} \cdot \frac{K_{O_2}}{K_{O_2}+S_{O_2}} \cdot \frac{K_{NO_3}}{K_{NO_3}+S_{NO_3}} \cdot \frac{S_A}{K_A+S_A} \cdot \frac{S_A}{S_F+S_A} \cdot \frac{S_{NH_4}}{K_{NH_4}+S_{NH_4}} \cdot \frac{S_{PO_4}}{K_P+S_{PO_4}} \cdot \frac{S_{ALK}}{K_{ALK}+S_{ALK}} \cdot X_H$
8	Fermentation	$q_{fe} \cdot \frac{K_{O_2}}{K_{O_2}+S_{O_2}} \cdot \frac{K_{NO_3}}{K_{NO_3}+S_{NO_3}} \cdot \frac{S_F}{K_F+S_F} \cdot \frac{S_{ALK}}{K_{ALK}+S_{ALK}} \cdot X_H$
9	Lysis	$b_H \cdot X_H$
Phosphorus-accumulating organisms (PAO): X_{PAO}		
10	Storage of X_{PHA}	$q_{PHA} \cdot \frac{S_A}{K_A+S_A} \cdot \frac{S_{ALK}}{K_{ALK}+S_{ALK}} \cdot \frac{X_{PP}/X_{PAO}}{K_{PP}+X_{PP}/X_{PAO}} \cdot X_{PAO}$
11	Aerobic storage of X_{PP}	$q_{PP} \cdot \frac{S_{O_2}}{K_{O_2}+S_{O_2}} \cdot \frac{S_{PO_4}}{K_{PS}+S_{PO_4}} \cdot \frac{S_{ALK}}{K_{ALK}+S_{ALK}} \cdot \frac{X_{PHA}/X_{PAO}}{K_{PHA}+X_{PHA}/X_{PAO}} \cdot \frac{K_{MAX}-X_{PP}/X_{PAO}}{K_{PP}+K_{MAX}-X_{PP}/X_{PAO}} \cdot X_{PAO}$
12	Anoxic storage of X_{PP}	$\rho_{12} = \rho_{11} \cdot \eta_{NO_3} \cdot \frac{K_{O_2}}{S_{O_2}} \cdot \frac{S_{NO_3}}{K_{NO_3}+S_{NO_3}}$
13	Aerobic growth on X_{PHA}	$\mu_{PAO} \cdot \frac{S_{O_2}}{K_{O_2}+S_{O_2}} \cdot \frac{S_{NH_4}}{K_{NH_4}+S_{NH_4}} \cdot \frac{S_{PO_4}}{K_P+S_{PO_4}} \cdot \frac{S_{ALK}}{K_{ALK}+S_{ALK}} \cdot \frac{X_{PHA}/X_{PAO}}{K_{PHA}+X_{PHA}/X_{PAO}} \cdot X_{PAO}$
14	Anoxic growth on X_{PP}	$\rho_{14} = \rho_{13} \cdot \eta_{NO_3} \cdot \frac{K_{O_2}}{S_{O_2}} \cdot \frac{S_{NO_3}}{K_{NO_3}+S_{NO_3}}$
15	Lysis of X_{PAO}	$b_{PAO} \cdot X_{PAO} \cdot S_{ALK}/(K_{ALK}+S_{ALK})$
16	Lysis of X_{PP}	$b_{PP} \cdot X_{PP} \cdot S_{ALK}/(K_{ALK}+S_{ALK})$
17	Lysis of X_{PHA}	$b_{PHA} \cdot X_{PHA} \cdot S_{ALK}/(K_{ALK}+S_{ALK})$
Nitrifying organisms (autotrophic organisms): X_{AUT}		
18	Aerobic growth of X_{AUT}	$\mu_{AUT} \cdot \frac{S_{O_2}}{K_{O_2}+S_{O_2}} \cdot \frac{S_{NH_4}}{K_{NH_4}+S_{NH_4}} \cdot \frac{S_{PO_4}}{K_P+S_{PO_4}} \cdot \frac{S_{ALK}}{K_{ALK}+S_{ALK}} \cdot X_{AUT}$
19	Lysis of X_{AUT}	$b_{AUT} \cdot X_{AUT}$
Simultaneous precipitation of phosphorus with ferric hydroxide $Fe(OH)_3$		
20	Precipitation	$k_{PRE} \cdot S_{PO_4} \cdot X_{MeOH}$
21	Redissolution	$k_{RED} \cdot X_{MeP} \cdot S_{ALK}/(K_{ALK}+S_{ALK})$

4. Typical wastewater characteristics and kinetic and stoichiometric constants

It is the responsibility of the user of the Activated Sludge Model No. 2 (ASM2 and ASM2d) to determine the concentrations of relevant components in the wastewater, as well as the stoichiometric and kinetic parameters which apply to the specific case to be dealt with. Absolute numbers of these parameters are neither part of ASM2 nor of ASM2d, but are necessary for the application of the model to a specific case.

In this section, the Task Group suggests a list of typical concentrations of model components in a primary effluent as well as a set of model parameters. This neither indicates that ASM2 or ASM2d is meant to be reliable with these parameters in any case, nor that these parameters are the state of the art. They are merely presented as a reference for testing computer code and a first estimate for the design of possible experiments which are proposed to determine these parameters more accurately.

Table 4.1 contains a list of all model components and typical concentrations in a primary effluent. This wastewater contains a total COD of 260 g COD m^{-3}, a total nitrogen content of 25 g N m^{-3} and approximately 140 g TSS m^{-3}. The analytically measured TSS are lower than the value of X_{TSS} = 180 g TSS m^{-3}, since a fraction of X_S in the influent would pass through membrane filters but must be included in the

Table 4.1 Short definition of model components and typical wastewater composition (primary effluent), considering the composition of the different model components as indicated in Table 4.2.

COD_{tot} = 260 g COD m^{-3}, TKN = 25 g N m^{-3}, TP = 6 g P m^{-3}			
Dissolved components:			
S_{O_2}	Dissolved oxygen	0	g O_2 m^{-3}
S_F	Readily biodegradable substrate	30	g COD m^{-3}
S_A	Fermentation products (acetate)	20	g COD m^{-3}
S_{NH_4}	Ammonium	16	g N m^{-3}
S_{NO_3}	Nitrate (plus nitrite)	0	g N m^{-3}
S_{PO_4}	Phosphate	3.6	g P m^{-3}
S_I	Inert, bon-biodegradable organics	30	g COD m^{-3}
S_{ALK}	Bicarbonate alkalinity	5	mole HCO_3^- m^{-3}
Particulate components:			
X_I	Inert, non-biodegradable organics	25	g COD m^{-3}
X_S	Slowly biodegradable substrate	125	g COD m^{-3}
X_H	Heterotrophic biomass	30	g COD m^{-3}
X_{PAO}	Phosphorus-accumulating organisms, PAO	0	g COD m^{-3}
X_{PP}	Stored poly-phosphate of PAO	0	g P m^{-3}
X_{PHA}	Organic storage products of PAO	0	g COD m^{-3}
X_{AUT}	Autotrophic, nitrifying biomass	0	g COD m^{-3}
X_{MeOH}	'Ferric-hydroxide', $Fe(OH)_3$	0	g $Fe(OH)_3$ m^{-3}
X_{MeP}	'Ferric-phosphate', $FePO_4$	0	g $FePO_4$ m^{-3}
X_{TSS}	Particulate material as model component[a)]	180[a)]	g TSS m^{-3}

a) This value is larger than TSS which may be measured analytically, since it includes the fraction of X_S, which would pass the filter in the TSS analysis. X_{TSS} may also include some inert mineral material, which is contained in the influent but not accounted for by other components. If this is the case, then X_{TSS} in the influent will be larger than predicted from the conservation equation, which for the above values and based on the conversion factors given in Table 4.2 would result in 140 g TSS m^{-3}. Analytically measured TSS (0.45 µm) would be approximately 120 g TSS m^{-3}.

Table 4.2. Definition and typical values for the stoichiometric coefficients of ASM2.

Typical conversion factors for conservation equation			
Nitrogen:			
Soluble material:			
i_{N,S_I}	N content of inert soluble COD S_I	0.01	g N (g COD)$^{-1}$
i_{N,S_F}	N content of fermentable substrates S_F	0.03	g N (g COD)$^{-1}$
Particulate material:			
i_{N,X_I}	N content of inert particulate COD X_I	0.02	g N (g COD)$^{-1}$
i_{N,X_S}	N content of slowly biodegradable substrate X_S	0.04	g N (g COD)$^{-1}$
$i_{N,BM}$	N content of biomass, X_H, X_{PAO}, X_{AUT}	0.07	g N (g COD)$^{-1}$
Phosphorus:			
Soluble material:			
i_{P,S_I}	P content of inert soluble COD S_I	0.00	g P (g COD)$^{-1}$
i_{P,S_F}	P content of fermentable substrates S_F	0.01	g P (g COD)$^{-1}$
Particulate material:			
i_{P,X_I}	P content of inert particulate COD X_I	0.01	g P (g COD)$^{-1}$
i_{P,X_F}	P content of slowly biodegradable substrate X_S	0.01	g P (g COD)$^{-1}$
$i_{P,BM}$	P content of biomass, X_H, X_{PAO}, X_{AUT}	0.02	g P (g COD)$^{-1}$
Total suspended solids: TSS			
i_{TSS,X_I}	TSS to COD ratio for X_I	0.75	g TSS (g COD)$^{-1}$
i_{TSS,X_S}	TSS to COD ratio for X_S	0.75	g TSS (g COD)$^{-1}$
$i_{TSS,BM}$	TSS to COD ratio for biomass, X_H, X_{PAO}, X_A	0.90	g TSS (g COD)$^{-1}$
Typical stoichiometric parameters			
Hydrolysis:			
f_{S_I}	Production of S_I in hydrolysis	0	g COD (g COD)$^{-1}$
Heterotrophic biomass: X_H			
Y_H	Yield coefficient	0.625	g COD (g COD)$^{-1}$
f_{X_I}	Fraction of inert COD generated in biomass lysis	0.10	g COD (g COD)$^{-1}$
Phosphorus-accumulating organisms: X_{PAO}			
Y_{PAO}	Yield coefficient (biomass/PHA)	0.625	g COD (g COD)$^{-1}$
Y_{PO_4}	PP requirement (PO_4 release) per PHA stored	0.40	g P (g COD)$^{-1}$
Y_{PHA}	PHA requirement for PP storage	0.20	g COD (g P)$^{-1}$
f_{X_I}	Fraction of inert COD generated in biomass lysis	0.10	g COD (g COD)$^{-1}$
Nitrifying organisms: X_{AUT}			
Y_A	Yield of autotrophic biomass per NO_3^--N	0.24	g COD (g N)$^{-1}$
f_{X_I}	Fraction of inert COD generated in biomass lysis	0.10	g COD (g COD)$^{-1}$

model component X_{TSS} since it will later adsorb onto the activated sludge. The total nitrogen (and phosphorus) in the influent may be computed with the aid of all influent concentrations multiplied with the relevant conversion factors from Tables 3.1 and 4.2.

Table 4.2 is a list of typical stoichiometric coefficients of ASM2 and ASM2d and includes the factors which are required for the use of the conservation equations (see also Table 3.1). Many of the conversion factors have been estimated without performing specific experiments for their determination. These values indicate an order of magnitude. The stoichiometric coefficients are either based on previous experience with ASM1 or they are derived from verification trials of ASM2 relative to full-scale experience. Experience with the three yield coefficients, Y_{PAO}, Y_{PO_4} and Y_{PHA} of the PAO are still scarce.

Table 4.3 is a summary of the definitions and typical values of all kinetic parameters of the models ASM2 and ASM2d. Again, some kinetic parameters were estimated based on the experience with ASM1, those relating to biological phosphorus removal are estimated based on laboratory experience and full-scale verification trials of ASM2. Note that saturation coefficients K_i for any specific compound may be different for different organisms (e.g. K_{O_2} may have four different values, depending on the process and organism to which it relates).

Table 4.3. Definition and typical values for the kinetic parameters of ASM2d.

Temperature:		20 °C	10 °C	Units
Hydrolysis of particulate substrate: X_S				
K_h	= Hydrolysis rate constant	3.00	2.00	d^{-1}
η_{NO_3}	= Anoxic hydrolysis reduction factor	0.60	0.60	–
η_{fe}	= Anaerobic hydrolysis reduction factor	0.40	0.40	–
K_{O_2}	= Saturation/inhibition coefficient for oxygen	0.20	0.20	$g\ O_2\ m^{-3}$
K_{NO_3}	= Saturation/inhibition coefficient for nitrate	0.50	0.50	$g\ N\ m^{-3}$
K_X	= Saturation coefficient for particulate COD	0.10	0.10	$g\ X_S\ (g\ X_H)^{-1}$
Heterotrophic organisms: X_H				
μ_H	= Maximum growth rate on substrate	6.00	3.00	$g\ X_S\ (g\ X_H)^{-1}\ d^{-1}$
q_{fe}	= Maximum rate for fermentation	3.00	1.50	$g\ S_F\ (g\ X_H)^{-1}\ d^{-1}$
η_{NO_3}	= Reduction factor for denitrification	0.80	0.80	–
b_H	= Rate constant for lysis and decay	0.40	0.20	d^{-1}
K_{O_2}	= Saturation/inhibition coefficient for oxygen	0.20	0.20	$g\ O_2\ m^{-3}$
K_F	= Saturation coefficient for growth on S_F	4.00	4.00	$g\ COD\ m^{-3}$
K_{fe}	= Saturation coefficient for fermentation of S_F	4.00	4.00	$g\ COD\ m^{-3}$
K_A	= Saturation coefficient for growth on acetate S_A	4.00	4.00	$g\ COD\ m^{-3}$
K_{NO_3}	= Saturation/inhibition coefficient for nitrate	0.50	0.50	$g\ N\ m^{-3}$
K_{NH_4}	= Saturation coefficient for ammonium (nutrient)	0.05	0.05	$g\ N\ m^{-3}$
K_P	= Saturation coefficient for phosphate (nutrient)	0.01	0.01	$g\ P\ m^{-3}$
K_{ALK}	= Saturation coefficient for alkalinity (HCO_3^-)	0.10	0.10	mole $HCO_3^-\ m^{-3}$
Phosphorus-accumulating organisms: X_{PAO}				
q_{PHA}	= Rate constant for storage of X_{PHA} (base X_{PP})	3.00	2.00	$g\ X_{PHA}\ (g\ X_{PAO})^{-1}\ d^{-1}$
q_{PP}	= Rate constant for storage of X_{PP}	1.50	1.00	$g\ X_{PP}\ (g\ X_{PAO})^{-1}\ d^{-1}$
μ_{PAO}	= Maximum growth rate of PAO	1.00	0.67	d^{-1}
η_{NO_3}	= Reduction factor for anoxic activity	0.60	0.60	–
b_{PAO}	= Rate for lysis of X_{PAO}	0.20	0.10	d^{-1}
b_{PP}	= Rate for lysis of X_{PP}	0.20	0.10	d^{-1}
b_{PHA}	= Rate for lysis of X_{PHA}	0.20	0.10	d^{-1}
K_{O_2}	= Saturation/inhibition coefficient for oxygen	0.20	0.20	$g\ O_2\ m^{-3}$
K_{NO_3}	= Saturation coefficient for nitrate, S_{NO_3}	0.50	0.50	$g\ N\ m^{-3}$
K_A	= Saturation coefficient for acetate, S_A	4.00	4.00	$g\ COD\ m^{-3}$
K_{NH_4}	= Saturation coefficient for ammonium (nutrient)	0.05	0.05	$g\ N\ m^{-3}$
K_{PS}	= Saturation coefficient for phosphorus in storage of PP	0.20	0.20	$g\ P\ m^{-3}$
K_P	= Saturation coefficient for phosphate (nutrient)	0.01	0.01	$g\ P\ m^{-3}$
K_{ALK}	= Saturation coefficient for alkalinity (HCO_3^-)	0.10	0.10	mole $HCO_3^-\ m^{-3}$
K_{PP}	= Saturation coefficient for poly-phosphate	0.01	0.01	$g\ X_{PP}\ (g\ X_{PAO})^{-1}$
K_{MAX}	= Maximum ratio of X_{PP}/X_{PAO}	0.34	0.34	$g\ X_{PP}\ (g\ X_{PAO})^{-1}$
K_{IPP}	= Inhibition coefficient for PP storage	0.02	0.02	$g\ X_{PP}\ (g\ X_{PAO})^{-1}$
K_{PHA}	= Saturation coefficient for PHA	0.01	0.01	$g\ X_{PHA}\ (g\ X_{PAO})^{-1}$
Nitrifying organisms (autotrophic organisms): X_{AUT}				
μ_{AUT}	= Maximum growth rate of X_{AUT}	1.00	0.35	d^{-1}
b_{AUT}	= Decay rate of X_{AUT}	0.15	0.05	d^{-1}
K_{O_2}	= Saturation coefficient for oxygen	0.50	0.50	$g\ O_2\ m^{-3}$
K_{NH_4}	= Saturation coefficient for ammonium (substrate)	1.00	1.00	$g\ N\ m^{-3}$
K_{ALK}	= Saturation coefficient for alkalinity (HCO_3^-)	0.50	0.50	mole $HCO_3^-\ m^{-3}$
K_P	= Saturation coefficient for phosphorus (nutrient)	0.01	0.01	$g\ P\ m^{-3}$
Precipitation:				
k_{PRE}	= Rate constant for P precipitaion	1.00	1.00	$m^3\ (g\ Fe(OH)_3)^{-1}\ d^{-1}$
k_{RED}	= Rate constant for redissolution	0.60	0.60	d^{-1}
K_{ALK}	= Saturation coefficient for alkalinity	0.50	0.50	mole $HCO_3^-\ m^{-3}$

Table 4.4. An example of a stoichiometric matrix for ASM2d for soluble and particulate components and for precipitation processes. The absolute values of the stoichiometric coefficients are based on the typical stoichiometric parameters introduced in Table 4.2. These values are not the ASM2d but rather a typical application of the model.

Stoichiometric matrix for soluble components

Process component	S_{O_2}	S_F	S_A	S_I	S_{NH_4}	S_{N_2}	S_{NO_3}	S_{PO_4}	S_{ALK}
expressed as →	O_2	COD	COD	COD	N	N	N	P	mole
1 Aerobic hydrolysis		1.00			0.01				0.001
2 Anoxic hydrolysis		1.00			0.01				0.001
3 Anaerobic hydrolysis		1.00			0.01				0.001
Heterotrophic organisms: X_H									
4 Growth on S_F	−0.60	−1.60			−0.022			−0.004	−0.001
5 Growth on S_A	−0.60		−1.60		−0.07			−0.02	0.021
6 Denitrification with S_F		−1.60			−0.022	0.21	−0.21	−0.004	0.014
7 Denitrification with S_A			−1.60		−0.07	0.21	−0.21	−0.02	0.036
8 Fermentation of S_F		−1	1		0.03			0.01	−0.014
9 Lysis					0.031			0.01	0.002
Phosphorus-accumulating organisms (PAO): X_{PAO}									
10 Storage of PHA			−1					0.40	0.009
11 Aerobic storage of PP	−0.20							−1	0.016
12 Anoxic storage of PP						0.07	−0.07	−1	0.021
13 Aerobic growth	−0.60				−0.07			−0.02	−0.004
14 Anoxic growth					−0.07	0.21	−0.21	−0.02	0.011
15 Lysis of PAO					0.031			0.01	0.002
16 Lysis of PP								1	−0.016
17 Lysis of PHA			1						−0.016
Nitrifying organisms (autotrophic organisms): X_{AUT}									
18 Aerobic growth	−18.0				−4.24		4.17	−0.02	−0.60
19 Lysis					0.031			0.01	0.002
Simultaneous precipitation of phosphorus with ferric hydroxide ($Fe(OH_3)$):									
20 Precipitation								−1	0.048
21 Redissolution								1	−0.048

Stoichiometric matrix for particulate components

Process component	X_I	X_S	X_H	X_{PAO}	X_{PP}	X_{PHA}	X_A	X_{TSS}	X_{MeOH}	X_{MeP}
expressed as →	COD	COD	COD	COD	P	COD	COD	TSS	TSS	TSS
1 Aerobic hydrolysis		−1						−0.75		
2 Anoxic hydrolysis		−1						−0.75		
3 Anaerobic hydrolysis		−1						−0.75		
Heterotrophic organisms: X_H										
4 Growth on S_S			1					0.90		
5 Growth on S_A			1					0.90		
6 Denitrification with S_S			1					0.90		
7 Denitrification with S_A			1					0.90		
8 Fermentation										
9 Lysis	0.10	0.9	−1					−0.15		
Phosphorus-accumulating organisms (PAO): X_{PAO}										
10 Storage of PHA					−0.40	1		−0.69		
11 Aerobic storage of PP					1	−0.20		3.11		
12 Anoxic storage of PP					1	−0.20		3.11		
13 Aerobic growth				1		−1.60		−0.06		
14 Anoxic growth				1		−1.60		−0.06		
15 Lysis of PAO	0.10	0.90		−1				−0.15		
16 Lysis of PP					−1			−3.23		
17 Lysis of PHA						−1		−0.60		
Nitrifying organisms (autotrophic organisms): X_{AUT}										
18 Aerobic growth							1	0.90		
19 Lysis	0.10	0.90					−1	−0.15		
Simultaneous precipitation of phosphorus with ferric hydroxide ($Fe(OH_3)$):										
20 Precipitation								1.42	−3.45	4.87
21 Redissolution								−1.42	3.45	−4.87

Future experience may well lead to different 'good estimates' of the parameters of the model. Since experimental results of many pilot studies have been performed without considering the requirements of model calibration, we do not currently have a sufficient basis to calibrate ASM2 or ASM2d to a 'typical wastewater'.

Finally a full stoichiometric matrix for ASM2d, based on the proposed stoichiometric parameters in Table 4.2 is presented in Table 4.4. Table 4.4 is not meant to be a part of ASM2d but rather it should indicate approximate values of stoichiometric coefficients $\nu_{j,i}$. Table 4.4 may be used to test computer code, which might be developed to predict stoichiometric coefficients $\nu_{j,i}$ based on conversion factors and stoichiometric constants as introduced in Table 4.2.

5. Limitations

All models have limitations. For ASM2d among the more important ones are:

- the model is valid for municipal wastewater only
- processes with overflow of S_A to the aeration tank cannot be modelled
- the wastewater must contain sufficient Mg^{2+} and K^+
- pH should be near neutral
- temperature is expected to be in the range of 10–25 °C

Use of the model outside of these limitations is not recommended.

6. Conclusion

ASM2d should be used as a basis for modelling of simultaneous biological phosphorus uptake and nitrification–denitrification. As compared with ASM2 it will improve the accuracy when modelling nitrate and phosphate dynamics. ASM2d is considered to be a platform and a reference for further research and development of kinetic models for biological nutrient removal in activated sludge systems.

7. References and bibliography

Henze, M., Gujer, W., Mino, T., Matsuo, T., Wentzel, M.C. and Marais, G.v.R. (1995) *Activated Sludge Model No. 2*. (IAWQ Scientific and Technical Report No. 3). London: IAWQ.

Henze, M., Grady, C.P.L. Jr, Gujer, W., Marais, G.v.R. and Matsuo, T. (1987) *Activated Sludge Model No. 1*. (IAWPRC Scientific and Technical Report No. 1). London: IAWPRC.

Kerrn-Jespersen, J.P. and Henze, M. (1993) Biological phosphorus uptake under anoxic and aerobic conditions. *Wat. Res.* **27**, 617–624.

Meinhold, J., Filipe, C.D.M., Daigger, G.D. and Isaacs, S. (1999) Characterization of the denitrifying fraction of phosphate accumulating organisms in biological phosphate removal. *Wat. Sci. Technol.* **39** (1), 31–42.

Mino, T., Liu, W-T., Kurisu, F. and Matsuo, T. (1995) Modelling glycogen storage and denitrification capability of microorganisms in enhanced biological phosphate removal processes. *Wat. Sci. Technol.* **31** (2), 25–34.

Bibliography

The following literature is not explicitly cited in this text but it has significantly contributed to the process of model building in the Task Group.

Brdjanovic, D (1998) Modeling biological phosphorus removal in activated sludge systems. Ph.D thesis, IHEE and Delft University of Technology.

Wanner, J., Čech, J.S. and Kos, M. (1992) New process design for biological nutrient removal. *Wat. Sci. Technol.* **25**, (4–5), 445–448.

ACTIVATED SLUDGE MODEL NO. 3

by

IAWQ TASK GROUP ON MATHEMATICAL MODELLING FOR DESIGN AND OPERATION OF BIOLOGICAL WASTEWATER TREATMENT

Willi Gujer, *Swiss Federal Institute of Technology, Switzerland*
Mogens Henze, *Technical University of Denmark, Denmark*
Takashi Mino, *Tokyo University, Japan*
Mark C.M. Van Loosdrecht, *Delft University of Technology, The Netherlands*

Contents

1. Introduction

With the introduction of the Activated Sludge Model No. 1 (ASM1) the IAWPRC (later IAWQ and now IWA) Task Group on Mathematical Modelling for Design and Operation of Biological Wastewater Treatment Processes introduced a new paradigm for the mathematical modelling of activated sludge systems. ASM1 as it was introduced in 1987 (Henze *et al.*, 1987) has become a major reference for many scientific an practical projects. Today, mathematical models related to ASM1 are implemented in various computer codes for the simulation of the behaviour of activated sludge systems treating municipal wastewater of mainly domestic origin.

With over ten years of experience with the application of ASM1, some defects of this model have become apparent, including:

- ASM1 does not include kinetic expressions that can deal with nitrogen and alkalinity limitations of heterotrophic organisms. The result is that computer code cannot be based on the original form of ASM1, where under some circumstances negative concentrations of, for example, ammonium may occur. This led to the development of computer codes based on different versions of ASM1, which can hardly be differentiated any more.
- ASM1 includes biodegradable soluble and particulate organic nitrogen as model compounds. These cannot easily be measured and made the use of ASM1 unnecessarily complicated. Therefore this distinction of nitrogen compounds has in the meantime been eliminated in many models based on ASM1.
- The kinetics of ammonification in ASM1 cannot easily be quantified, moreover the process is fast and therefore hardly affects model predictions. Again in many versions of ASM1 assuming a constant composition of all organic compounds (constant N to COD ratio) has eliminated this process.
- ASM1 differentiates inert particulate organic material depending on its origin, influent or biomass decay, but it is impossible to differentiate these two fractions in reality.
- In the structure of ASM1, the process of hydrolysis has a dominating effect upon the predictions of oxygen consumption and denitrification by heterotrophic organisms. In reality this process stands for some coupled processes such as hydrolysis, lysis of organisms and storage of substrates. Therefore the identification of the kinetic parameters for this combined process is difficult.
- Lysis combined with hydrolysis and growth is used to describe the lumped effects of endogenous respiration of, for example, storage compounds, death, predation and lysis of the biomass. This leads to further difficulties in the evaluation of kinetic parameters.
- With elevated concentrations of readily biodegradable organic substrates, storage of poly-hydroxy-alkanoates and sometimes lipids or glycogen is observed under aerobic and anoxic conditions in activated sludge plants. This process is not included in ASM1.
- ASM1 does not include the possibility to differentiate decay rates of nitrifiers under aerobic and anoxic conditions. At high solids retention times (SRT) and high fractions of anoxic reactor volumes this leads to problems with the prediction of maximum nitrification rates.
- ASM1 does not directly predict the frequently measured mixed liquor suspended solids concentration.
- In respiration tests frequently high biomass yield coefficients are obtained. Even if only soluble, readily biodegradable substrates such as acetate are added, it appears from respiration tests that this substrate includes a slowly biodegradable fraction.

Considering all these defects and the advance in experimental evidence on storage of organic compounds, the Task Group has proposed the Activated Sludge Model No. 3 (ASM3) (Gujer *et al.*, 1999) which should correct for these defects and which could become a new standard for future modelling. ASM3 relates to the same dominating phenomena as does ASM1: oxygen consumption, sludge production, nitrification and denitrification in activated sludge systems treating wastewater of primarily domestic origin.

ASM3 is designed to be the **core** of many different models. Modules for biological phosphorus removal (as contained in the Activated Sludge Model No. 2 (ASM2 and ASM2d) (Henze *et al.*, 1995, 1999), chemical precipitation, growth of filamentous organisms or pH calculations are not part of ASM3 but can easily be connected as add on modules. With increasing experience with ASM3 the Task Group may well suggest such modules which would serve many purposes in practical simulation work.

Introduction of ASM1 has spurred and focused research internationally. Based on a common platform it became possible to discuss rather complex results of careful research. Today interest is with topics such as modelling population dynamics, biological phosphorus removal and structured biomass (storage products). ASM3 may provide the backbone, which describes the processes of minor interest in research, such that we can concentrate on new frontiers again. In this we should realize that scientific research and model application in engineering practice have different goals. Whereas the detailed structure of the models is used to convey the message on new mechanisms which have been identified in our advanced research projects, model application in engineering must rely on manageable models with a moderate number of parameters but a high potential to predict system behaviour. ASM3 is designed to satisfy primarily the requirements of model application. Nevertheless, the Task Group has tried to fulfil the didactic requirement to keep as many details as are necessary to obtain some insight into the interconnected processes. ASM3 may well become a basis for teaching advanced biological wastewater treatment courses.

2. Comparison of ASM1 and ASM3

In ASM1 a single decay process (lysis) was introduced to describe the sum of all decay processes under all environmental conditions (aerobic, anoxic). The reason was that in 1985, when ASM1 was first published, computing power was still scarce. The simplest description possible saved computation time. Today, as computation is not limiting simulation to the same extent, a more realistic description of decay processes is introduced in ASM3: endogenous respiration is close to the phenomena observed (we typically measure a respiration rate) and the relevant rate constants can be obtained directly and independent of stoichiometric parameters (from the slope of $\ln(r_{O_2,endog})$ versus time).

The flow of COD in ASM1 is rather complex. The death (decay) regeneration cycle of the heterotrophs and the decay process of nitrifiers are strongly interrelated (Figure 2.1). The two decay processes differ significantly in their details. This results in differing and confusing meanings of the two decay rates in ASM1. In ASM3 all the conversion processes of the two groups of organisms are clearly separated and decay processes are described with identical models (Figure 2.1).

The complexity of ASM3 is comparable to ASM1. There is a shift of emphasis from hydrolysis to storage of organic substrates, a process, which has been postulated and observed by many researchers. Characterization of wastewater must consider this change. Readily available organic substrates (S_S) must now be estimated based on the storage rather than the growth process. Differentiation of soluble and particulate substrates (S_S and X_S) remains somewhat arbitrary as in ASM1 and is mainly based on time constants for degradation. Correct characterization of wastewater for the use of ASM3 might still rely on bioassays, which relate to respiration.

Similarly to ASM2 (Henze *et al.* 1995) ASM3 includes cell internal storage compounds. This requires the biomass to be modelled with cell internal structure. Decay processes (which include predation) must include both fractions of the biomass, hence four decay processes are required (aerobic and anoxic loss of X_H as well as X_{STO}) and the kinetics of the growth processes (aerobic and anoxic) must relate to the ratio of X_{STO}/X_H.

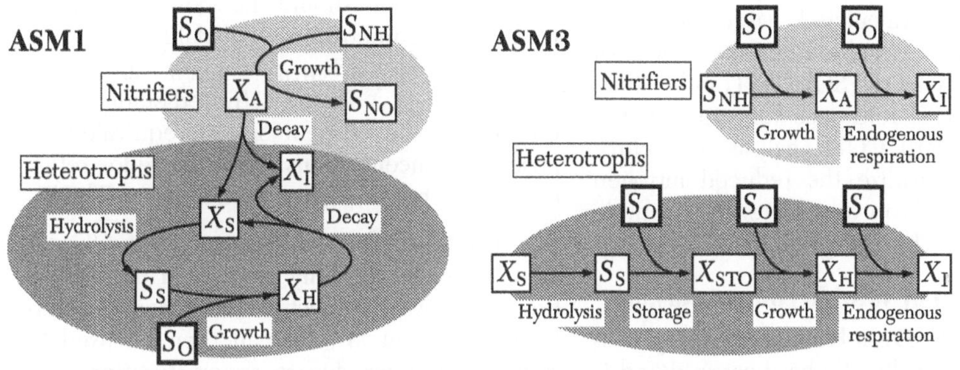

Figure 2.1. Flow of COD in ASM1 and ASM3. In ASM1 (left) heterotrophic organisms use COD in a cyclic reaction scheme: Decay feeds into hydrolysis and triggers additional growth. Nitrifiers decay and thereby enhance heterotrophic growth. Autotrophic and heterotrophic organisms cannot be entirely separated. Only two entry points for oxygen exist. In ASM3 (right) nitrifiers and heterotrophs are clearly separated, no COD flows from one group to the other. Many entry points for oxygen exist. For definitions of state variables see later.

3. ASM3: definition of compounds in the model

The following compounds are used in ASM3. Concentrations of soluble compounds are characterized by *S* and particulate compounds by *X*. Within the activated sludge systems the particulate compounds are assumed to be associated with the activated sludge (flocculated onto the activated sludge or contained within the active biomass). Particulate compounds can be concentrated by sedimentation/thickening in clarifiers whereas soluble compounds can only be transported with the water. Only soluble compounds may carry ionic charge. As in ASM1 and ASM2 filtration over 0.45 µm membrane filters cannot be used to differentiate model soluble from model particulate compounds in the influent (typically primary effluent): particulate, slowly biodegradable substrates (X_S, see later) will partially not be retained on the filter membrane. In the activated sludge reactors, where large amounts of surfaces exist, these substrates will rapidly adsorb to the suspended solids, resulting in a better differentiation of soluble and particulate compounds.

Conservation of **Theoretical Oxygen Demand** (ThOD) will be used extensively in the development of process stoichiometry. For organic materials COD may analytically approximate this ThOD. For some inorganic materials ThOD must be calculated based on redox equations relative to the redox reference of H_2O, CO_2, NH_4^+, PO_4^{3-}, SO_4^{2-}. NH_4 rather than NO_3^- is chosen as reference for nitrogen because the standard COD analysis with chromate does not oxidize the reduced nitrogen compounds present in wastewater. An example of calculation of ThOD is given in Table 3.1.

3.1. Definition of soluble compounds, $S_?$

S_{O_2} [M(O_2) L^{-3}]: Dissolved oxygen, O_2. Dissolved oxygen can directly be measured and is subject to gas exchange. In stoichiometric computations S_{O_2} is introduced as negative ThOD.

S_I [M(COD) L^{-3}]: Inert soluble organic material. The prime characteristic of S_I is that these organics cannot be further degraded in the treatment plants dealt with in this report. This material is assumed to be part of the influent and may be produced in the context of hydrolysis of particulate substrates X_S. It can readily be estimated from the residual soluble COD in the effluent of a low loaded activated sludge plant.

S_S [M(COD) L^{-3}]: Readily biodegradable organic substrates (COD). This fraction of the soluble COD is directly available for consumption by heterotrophic organisms. In ASM3, for simplification, it is assumed that all these substrates are first taken up by heterotrophic organisms and stored in the form of X_{STO}. S_S is preferentially determined with the aid of a bioassay (respiration test). Measuring the sum of $S_I + S_S$ in the form of the total soluble COD in wastewater as determined with 0.45 µm membrane filtration may lead to gross errors. This is due to the fact that some X_S (see later) in wastewater (e.g. starch) cannot adsorb to the small amount of biomass present in the influent and therefore contributes to the analytically determined soluble material.

S_{NH_4} [M(N) L^{-3}]: Ammonium plus ammonia nitrogen (NH_4^+-N + NH_3-N). For the balance of the ionic charges, S_{NH_4} is assumed to be all NH_4^+. Because ASM3 assumes that organic compounds contain a fixed fraction of organic nitrogen ($\iota_{N,i}$, see Table 8.2), the influent $S_{NH_4,0}$ cannot be observed directly (measured analytically) but should be computed from wastewater composition: Kjeldahl nitrogen - organic nitrogen ($S_{NH_4,0} = C_{TKN,0} - \Sigma\iota_{N,i} \cdot C_{i,0} + S_{N_2,0} + S_{NOX,0}$). In the activated sludge reactors and in the effluent S_{NH_4} is equivalent to observed concentrations. With the redox reference level chosen, S_{NH_4} does not have a ThOD.

S_{N_2} [M(N) L^{-3}]: Dinitrogen (N_2). S_{N_2} is assumed to be the only product of denitrification. S_{N_2} may be subject to gas exchange, parallel with oxygen, S_{O_2}. It can then be used to predict problems due to supersaturation with N_2 in secondary clarifiers. Alternatively the N_2 contained in the influent and gas exchange can be neglected. S_{N_2} may then be used to calculate the amount of nitrogen lost due to denitrification. S_{N_2} has a negative ThOD.

S_{NOX} [M(N) L^{-3}]: Nitrate plus nitrite nitrogen (NO_3^--N + NO_2^--N). S_{NOX} is assumed to include nitrate as well as nitrite nitrogen, since nitrite is not included as a separate model

Table 3.1. Computation of Theoretical Oxygen Demand ThOD. Each reactive electron is equivalent to a ThOD of 8 g mole^{-1}. Therefore each element can be associated with a ThOD which relates to the redox reference of ThOD (H_2O, CO_2, NH_4^+, SO_4^{2-}, PO_4^{3-}). ThOD may then be computed by adding the individual contributions to each molecule.

Element		Equivalent ThOD	Examples		
Carbon	C	+ 32 g ThOD (mol C)$^{-1}$	*What is the ThOD of 1 mole of NO_3^-?*		
Nitrogen	N	– 24 g ThOD (mol N)$^{-1}$	N:	– 24 g mole^{-1}	
Hydrogen	H	+ 8 g ThOD (mol H)$^{-1}$	3 O:	– 48	
Oxygen	O	– 16 g ThOD (mol O)$^{-1}$	–:	+ 8 g mole^{-1}	Total: –64 g ThOD (mole NO_3^-)$^{-1}$
Sulphur	S	+ 48 g ThOD (mol S)$^{-1}$	*What is the ThOD of 1 mole of SO_4^{2-}?*		
Phosphorus	P	+ 40 g ThOD (mol P)$^{-1}$	S:	+ 48 g mole^{-1}	
Negative charge	–	+ 8 g ThOD (mol (–))$^{-1}$	4 O:	– 64 g mole^{-1}	
Positive charge	+	– 8 g ThOD (mol (+))$^{-1}$	2–:	+ 16 g mole^{-1}	Total: 0 g ThOD (mole SO_4^{2-})$^{-1}$

compound. For all stoichiometric computations (ThOD conservation), S_{NOX} is considered to be NO_3^--N only. S_{NOX} has a negative ThOD.

S_{ALK} [mole(HCO_3) L^{-3}]: Alkalinity of the wastewater. Alkalinity is used to approximate the conservation of ionic charge in biological reactions. Alkalinity is introduced in order to obtain an early indication of possible low pH conditions, which might inhibit some biological processes. For all stoichiometric computations, S_{ALK} is assumed to be bicarbonate, HCO_3, only.

3.2. Definition of particulate compounds, $X_?$

X_I [M(COD) L^{-3}]: Inert particulate organic material (COD). This material is not degraded in the activated sludge systems for which ASM3 has been developed. It is flocculated onto the activated sludge. X_I may be a fraction of the influent and is produced in the context of biomass decay.

X_S [M(COD) L^{-3}]: Slowly biodegradable substrates (COD). Slowly biodegradable substrates are high molecular weight, soluble, colloidal and particulate organic substrates which must undergo cell external hydrolysis before they are available for degradation. It is assumed that the products of hydrolysis of X_S are either readily biodegradable (S_S) or inert (S_I) soluble organics. As compared to ASM1 this fraction has a different origin. In ASM3 all X_S is contained in the influent and none is generated in decay processes. In ASM1 a large fraction of X_S is assumed to originate from decay processes.

X_H [M(COD) L^{-3}]: Heterotrophic organisms (COD). These organisms are assumed to be the 'allrounder' heterotrophic organisms, they can grow aerobically and many of them also anoxically (denitrification). These organisms are responsible for hydrolysis of particulate substrates X_S and can metabolize all degradable organic substrates. They can form organic storage products in the form of poly-hydroxy-alkanoates or glycogen. X_H are assumed to have no anaerobic activity except cell external hydrolysis, which is the only anaerobic process in ASM3.

X_{STO} [M(COD) L^{-3}]: A cell internal storage product of heterotrophic organisms (COD). It includes poly-hydroxy-alkanoates (PHA), glycogen, etc. It occurs only associated with X_H; it is, however, not included in the mass of X_H. X_{STO} cannot be directly compared with analytically measured PHA or glycogen concentrations; X_{STO} is only a functional compound required for modelling but not directly identifiable chemically. X_{STO} may, however, be recovered in COD analysis and must satisfy ThOD conservation. For stoichiometric considerations, X_{STO} is assumed to have the chemical composition of poly-hydroxy-butyrate $(C_4H_6O_2)_n$.

X_A [M(COD) L^{-3}]: Nitrifying organisms (COD). Nitrifying organisms are responsible for nitrification; they are obligate aerobic, chemo-litho-autotrophic. It is assumed that nitrifiers oxidize ammonium, S_{NH_4}, directly to nitrate, S_{NOX}. Nitrite as an intermediate compound of nitrification is not considered in ASM3.

X_{SS} [M(SS) L^{-3}]: Suspended solids (SS). Suspended solids are introduced into the biokinetic models in order to compute their concentration via stoichiometry. Treatment plant operators typically follow SS in day to day analysis. In the influent, SS ($X_{SS,0}$) include an inorganic fraction of SS and the 'soluble' fraction of $X_{S,0}$, which passes membrane filters. SS measured in the influent are therefore smaller than $X_{SS,0}$ used to describe the influent in the terms of the model compounds. Describing influent SS correctly should allow predicting MLSS as observed in the activated sludge reactors. If chemicals are added in order to precipitate phosphorus, the precipitates formed must in ASM3 be added to the concentration of SS computed in the influent ($X_{SS,0}$). Alternatively X_{SS} may be used to model volatile suspended Solids (VSS). This requires the relevant choice of absolute numbers for the composition parameters for SS ($i_{SS,?}$ in Table 8.2).

4. ASM3: definition of processes in the model

ASM3 includes only the microbiological transformation processes. Chemical precipitation processes are not included, but may easily be added based on the information provided for ASM2 (Henze *et al.*, 1995). ASM3 considers the following transformation processes:

1. Hydrolysis. This process makes available all slowly biodegradable substrates X_S contained in the influent to an activated sludge system. Hydrolysis is assumed to be active independently of the electron donor. This process is different from the hydrolysis process in ASM1; it is of less dominating importance for the rates of oxygen consumption and denitrification.
2. Aerobic storage of readily biodegradable substrate. This process describes the storage of readily biodegradable substrate S_S in the form of cell internal storage products X_{STO}. This process requires energy, which is obtained from aerobic respiration. It is assumed that all substrates first become stored material and later are assimilated to biomass. This is definitely not observed in reality, however at this moment no reliable model is available which can predict the substrate flux into storage, assimilation and dissimilation respectively. Therefore the Task Group suggests for the time being this simplest assumption. However using a low yield coefficient for storage (Y_{STO}) and a higher one for subsequent growth (Y_H) allows to approximate the consequences of direct growth rather than storage followed by growth.
3. Anoxic storage of readily biodegradable substrate. This process is identical to aerobic storage, but denitrification rather than aerobic respiration provides the energy required. Only a fraction of the heterotrophic organisms X_H in activated sludge is capable of denitrification. ASM3 considers this by reducing the anoxic heterotrophic storage rate as compared to the aerobic rate.
4. Aerobic growth of heterotrophs. The substrate for the growth of heterotrophic organisms is assumed to consist entirely of stored organics X_{STO}. This assumption simplifies ASM3 considerably.
5. Anoxic growth of heterotrophs. This process is similar to aerobic growth but respiration is based on denitrification. Only a fraction of the heterotrophic organisms X_H in activated sludge is capable of denitrification. ASM3 considers this by reducing the anoxic heterotrophic storage rate as compared to the aerobic rate.
6. Aerobic endogenous respiration. This process describes all forms of biomass loss and energy requirements not associated with growth by considering related respiration under aerobic conditions: decay, (maintenance), endogenous respiration, lysis, predation, motility, death, and so on. The model of this process is significantly different from the decay (lysis) process introduced in ASM1.
7. Anoxic endogenous respiration. This process is similar to aerobic endogenous respiration but typically slower. Especially protozoa (predation) are considerably less active under denitrifying than under aerobic conditions.
8. Aerobic respiration of storage products. This process is analogous to endogenous respiration. It assures that storage products, X_{STO}, decay together with biomass.
9. Anoxic respiration of storage products. This process is analogous to the aerobic process but under denitrifying conditions.

As compared with ASM1, ASM3 includes a more detailed description of cell internal processes (storage) and allows for better adjustment of decay processes to environmental conditions. The importance of hydrolysis has been reduced and degradation of soluble and particulate organic nitrogen have been integrated into the hydrolysis, decay and growth process.

5. ASM3: stoichiometry

Table 5.1 introduces the stoichiometric matrix $\nu_{j,i}$ of ASM3 together with the composition matrix $\iota_{k,i}$ as proposed by Gujer and Larsen (1995). Whereas the stoichiometric matrix $\nu_{j,i}$ is well known since the introduction of ASM1, the composition matrix is less well known. Relating to Table 5.1 the composition matrix may be read as follows: $\iota_{2,3}$ is filled with the symbol $i_{N,SS}$ and indicates that any g COD in the form of S_S contains $i_{N,SS}$ g of N. The index $k = 2$ relates to the second conservative which is nitrogen, the index $i = 3$ relates to the third compound which is S_S. S_S is measured in terms of g COD (as indicated below the symbol S_S) and the conservative 'nitrogen' is expressed in g N (as indicated to the right of nitrogen in the composition matrix). $i_{N,SS}$ therefore indicates the composition of S_S relative to nitrogen, hence $\iota_{k,i}$ is called the composition matrix.

All empty elements of $\nu_{j,i}$ or $\iota_{k,i}$ indicate values of 0. All values of x_j, y_j and z_j can be obtained from the conservation Equation 5.1 for the three conservatives k: ThOD, nitrogen and ionic charge:

$$\sum_i \nu_{j,i} \cdot \iota_{k,i} = 0 \quad \text{for } i = 1 \text{ to } 12 \qquad (5.1)$$

As introduced earlier, ThOD stands for Theoretical Oxygen Demand and is the conservative form of COD. In most cases ThOD of organic compounds may analytically be approximated by standard dichromate COD analysis. ThOD is a conservative quantity since it effectively accounts for the electrons involved in the biological redox processes.

The stoichiometric coefficient for S_{N_2} in any denitrification process is the negative of the coefficient for S_{NOX}. The composition coefficients for ThOD for S_{N_2} (-1.71 g ThOD (g $N_2)^{-1}$) and S_{NOX} (-4.57 g ThOD (g NO_3^--N$)^{-1}$) as well as S_{O_2} (-1 g ThOD (g $O_2)^{-1}$) are negative for electron donors relative to the redox reference for ThOD.

The stoichiometric coefficients for the observable X_{SS} can be obtained from the Composition Equation 5.2:

$$\nu_{j,13} = \sum_i \nu_{j,i} \cdot \iota_{4,i} \quad \text{for } i = 8 \text{ to } 12 \qquad (5.2)$$

It is known that the biochemical energy (ATP) yield of anoxic respiration is smaller than in aerobic respiration. This leads to the fact that aerobic yield coefficients (Y_{STO,O_2} and Y_{H,O_2}) exceed the anoxic yield coefficients ($Y_{STO,NOX}$ and $Y_{H,NOX}$). Assuming the anoxic energy yield to be $\eta_{anoxic} = 0.70$ of the aerobic energy yield the following energy relationship (Equation 5.3) applies:

$$\frac{1 - Y_{STO,O_2}}{Y_{STO,O_2}} = \frac{\eta_{anoxic} \cdot (1 - Y_{STO,NOX})}{Y_{STO,NOX}} \quad \text{and}$$
$$\frac{1 - Y_{O_2}}{Y_{O_2}} = \frac{\eta_{anoxic} \cdot (1 - Y_{NOX})}{Y_{NOX}} \qquad (5.3)$$

It is suggested that Equation 5.3 is used to relate anoxic and aerobic yields in ASM3.

The net (true) yield of heterotrophic biomass X_H produced per unit of substrate S_S removed in ASM3 is obtained from:

$$Y_{net,O_2} = Y_{STO,O_2} \cdot Y_{H,O_2} \quad \text{and}$$
$$Y_{net,NOX} = Y_{STO,NOX} \cdot Y_{H,NOX} \qquad (5.4)$$

All stoichiometric parameters are defined together with their units and a typical value in Table 8.2. A numeric example of all stoichiometric coefficients is given in Table 8.4.

In the composition matrix $\iota_{k,i}$ of Table 5.1 the composition of all organic fractions relative to ThOD is assumed to be unity. These values have units however (i_{ThOD,S_I} = 1 g ThOD (g COD$)^{-1}$, ...) and it should be realized that these values are actually model parameters which here have been assumed to be unity whereas in reality COD analysis recovers only a fraction of ThOD, typically 95% in domestic wastewater.

Table 5.1. Stoichiometric matrix $\nu_{j,i}$ and composition matrix $\iota_{k,i}$ of ASM3. The values of x_j, y_j, z_j and t_j can be obtained in this sequence from mass and charge conservation (Equation 5.1) and composition (Equation 5.2).

	Compound i →		1	2	3	4	5	6	7	8	9	10	11	12	13
j	Process		S_{O_2}	S_I	S_S	S_{NH_4}	S_{N_2}	S_{NOX}	S_{ALK}	X_I	X_S	X_H	X_{STO}	X_A	X_{SS}
↓	Expressed as →		O_2	COD	COD	N	N	N	Mole	COD	COD	COD	COD	COD	SS
1	Hydrolysis			f_{S_I}	x_1	y_1			z_1		−1				$-i_{X_S}$
	Heterotrophic organisms, aerobic and denitrifying activity														
2	Aerobic storage of S_S		x_2		−1	y_2			z_2				Y_{STO,O_2}		t_2
3	Anoxic storage of S_S				−1	y_3	$-x_3$	x_3	z_3				$Y_{STO,NOX}$		t_3
4	Aerobic growth of X_H		x_4			y_4			z_4			1	$-1/Y_{H,O_2}$		t_4
5	Anoxic growth (denitrific.)					y_4	$-x_5$	x_5	z_5			1	$-1/Y_{H,NOX}$		t_5
6	Aerobic endog. respiration		x_6			y_6			z_6	f_I		−1			t_6
7	Anoxic endog. respiration					y_7	$-x_7$	x_7	z_7	f_I		−1			t_7
8	Aerobic respiration of X_{STO}		x_8										−1		t_8
9	Anoxic respiration of X_{STO}						$-x_9$	x_9	z_9				−1		t_9
	Autotrophic organisms, nitrifying activity														
10	Aerobic growth of X_A		x_{10}			y_{10}		$1/Y_A$	z_{10}					1	t_{10}
11	Aerobic endog. respiration		x_{11}			y_{11}			z_{11}	f_I				−1	t_{11}
12	Anoxic endog. respiration					y_{12}	$-x_{12}$	x_{12}	z_{12}	f_I				−1	t_{12}
	Composition matrix $\iota_{k,I}$														
k	*Conservatives*														
1	ThOD	g ThOD	−1	1	1		−1.71	−4.57		1	1	1	1	1	
2	Nitrogen	g N		i_{N,S_I}	i_{N,S_S}	1	1	1		i_{N,X_I}	i_{N,X_S}	$i_{N,BM}$		$i_{N,BM}$	
3	Ionic charge	Mole +				1/14		−1/14	−1						
	Observables														
4	SS	g SS								i_{SS,X_I}	i_{SS,X_S}	$i_{SS,BM}$	0.60	$i_{SS,BM}$	

6. ASM3: kinetics

The kinetic expressions of ASM3 are based on switching functions (hyperbolic or saturation terms, Monod equations, $S/(K+S)$) for all soluble compounds consumed. This form of kinetic expression is chosen not because of experimental evidence but rather for mathematical convenience: these switching functions stop all biological activity as educts of a process approach zero concentrations, an important difference between ASM1 and ASM3. Similarly for particulate educts the switching functions relate to the ratio of X_{STO}/X_H resp. X_S/X_H. Inhibition is modelled with $1 - S/(K + S) = K/(K + S)$.

Table 6.1 is a summary of all kinetic expressions of ASM3. The kinetic parameters are defined in Table 8.1 together with their units and a typical value at 10 °C and 20 °C. It is recommended to interpolate kinetic parameters k to different temperatures T (in °C) with the following temperature equation:

$$k(T) = k(20\,°\mathrm{C}) \cdot \exp(\theta_T \cdot (T - 20\,°\mathrm{C})) \quad (6.1)$$

where θ_T (in °C) may be obtained from

$$\theta_T = \frac{\ln(k(T_1)/k(T_2))}{t_1 - T_2} \quad (6.2)$$

6.1 Estimation of readily biodegradable substrate S_S

As indicated earlier, readily biodegradable substrate S_S in wastewater is best determined from a bioassay. Kinetics of the storage and growth processes are such that the storage process in ASM3 will be related to the additional rapid uptake of oxygen after addition of wastewater to biomass. Therefore the yield of the storage

Table 6.1. Kinetic rate expressions ρ_j for ASM3. All $\rho_j \geq 0$.

j	Process	Process rate equation ρ_j, all $\rho_j \geq 0$.
1	Hydrolysis	$k_H \cdot \frac{X_S/X_H}{K_X + X_S/X_H} \cdot X_H$
	Heterotrophic organisms, aerobic and denitrifying activity	
2	Aerobic storage of S_S	$k_{STO} \cdot \frac{S_{O_2}}{K_{O_2} + S_{O_2}} \cdot \frac{S_S}{K_S + S_S} \cdot X_H$
3	Anoxic storage of S_S	$k_{STO} \cdot \eta_{NOX} \cdot \frac{K_{O_2}}{K_{O_2} + S_{O_2}} \cdot \frac{S_{NOX}}{K_{NOX} + S_{NOX}} \cdot \frac{S_S}{K_S + S_S} \cdot X_H$
4	Aerobic growth	$\mu_H \cdot \frac{S_{O_2}}{K_{O_2} + S_{O_2}} \cdot \frac{S_{NH_4}}{K_{NH_4} + S_{NH_4}} \cdot \frac{S_{ALK}}{K_{ALK} + S_{ALK}} \cdot \frac{X_{STO}/X_H}{K_{STO} + X_{STO}/X_H} \cdot X_H$
5	Anoxic growth (denitrification)	$\mu_H \cdot \eta_{NOX} \cdot \frac{K_{O_2}}{K_{O_2} + S_{O_2}} \cdot \frac{S_{NOX}}{K_{NOX} + S_{NOX}} \cdot \frac{S_{NH_4}}{K_{NH_4} + S_{NH_4}} \cdot \frac{S_{ALK}}{K_{ALK} + S_{ALK}} \cdot \frac{X_{STO}/X_H}{K_{STO} + X_{STO}/X_H} \cdot X_H$
6	Aerobic endogenous respiration	$b_{H,O_2} \cdot \frac{S_{O_2}}{K_{O_2} + S_{O_2}} \cdot X_H$
7	Anoxic endogenous respiration	$b_{H,NOX} \cdot \frac{K_{O_2}}{K_{O_2} + S_{O_2}} \cdot \frac{S_{NOX}}{K_{NOX} + S_{NOX}} \cdot X_H$
8	Aerobic respiration of X_{STO}	$b_{STO,O_2} \cdot \frac{S_{O_2}}{K_{O_2} + S_{O_2}} \cdot X_{STO}$
9	Anoxic respiration of X_{STO}	$b_{STO,NOX} \cdot \frac{K_{O_2}}{K_{O_2} + S_{O_2}} \cdot \frac{S_{NOX}}{K_{NOX} + S_{NOX}} \cdot X_{STO}$
	Autotrophic organisms, nitrifying activity	
10	Aerobic growth of X_A, nitrification	$\mu_A \cdot \frac{S_{O_2}}{K_{A,O_2} + S_{O_2}} \cdot \frac{S_{NH_4}}{K_{A,NH_4} + S_{NH_4}} \cdot \frac{S_{ALK}}{K_{A,ALK} + S_{ALK}} \cdot X_A$
11	Aerobic endogenous respiration	$b_{A,O_2} \cdot \frac{S_{O_2}}{K_{A,O_2} + S_{O_2}} \cdot X_A$
12	Anoxic endogenous respiration	$b_{A,NOX} \cdot \frac{K_{A,O_2}}{K_{A,O_2} + S_{O_2}} \cdot \frac{S_{NOX}}{K_{A,NOX} + S_{NOX}} \cdot X_A$

process must be used in order to relate oxygen uptake to substrate consumption:

$$S_S\,(\text{batch}) = \int \Delta r_{S_S} \cdot dt = \frac{\nu_{S_S}}{\nu_{SO_2}} \int \Delta r_{SO_2} \cdot dt = \frac{\int \Delta r_{SO_2} \cdot dt}{1 - Y_{O_2,STO}} \tag{6.3}$$

It is recommended to simulate the batch experiment, which is used to identify S_S with the aid of ASM3. This allows the identification of possible errors that could be introduced by this simple procedure.

7. Limitations of ASM3

ASM3 (and ASM1) was developed for the simulation of the aerobic and anoxic treatment of domestic wastewater in activated sludge systems. It is not advised to apply it to situations where industrial contributions dominate the characteristics of the wastewater.

ASM3 (and ASM1) has been developed based on experience in the temperature range of 8–23 °C. Outside of this range model application may lead to very significant errors and even model structure may become unsatisfactory.

ASM3 (and ASM1) does not include any processes that describe biomass behaviour in an anaerobic environment. Simulation of systems with large fractions of anaerobic reactor volume may therefore lead to gross errors.

Development of ASM3 is based on experience in the range of pH values from 6.5 to 7.5. The concentration of bicarbonate alkalinity (S_{ALK}) is supplied to give early warnings when pH values below this range are to be expected. Alkalinity must be dominated by bicarbonate.

ASM3 cannot deal with elevated concentrations of nitrite.

ASM3 (and ASM1) is not designed to deal with activated sludge systems with very high load or small SRT (<1 day) where flocculation/adsorption of X_S and storage may become limiting.

ASM3 provides the structure of a model but not absolute values of model parameters. It is the responsibility of the user of this model to identify the applicable parameters and the relevant characterization of the wastewater.

Neither the Task Group nor IWA can under any circumstances accept any liability for damages of any sort that may result from the application of this model. It is provided here as a service for the scientific and practical engineering community and it is hoped to serve as a reference for future scientific work.

8. Aspects of application of ASM3

It is the responsibility of the user of ASM3 to determine the concentrations of relevant compounds in the wastewater, as well as the stoichiometric and kinetic parameters, which apply to the specific case to be dealt with. **Absolute values of these parameters are not part of ASM3.** They are necessary, however, if ASM3 is to be applied to any specific case.

In Tables 8.1–8.4 a set of typical model parameters and concentrations of model compounds in a primary effluent is provided for convenience. This neither indicates that ASM3 is meant to be reliable with these parameters in any case, nor that these parameters are the state of the art. They are merely presented here as a reference for testing computer code and as a first estimate for the design of possible experiments that may be used to identify these parameters more accurately.

Table 8.1 contains a list of typical kinetic parameters; Table 8.2 suggests some typical stoichiometric parameters. Table 8.3 indicates the composition of a typical primary effluent and finally Table 8.4 is a stoichiometric matrix, based on Table 5.1 and the specific values introduced in Table 8.2.

Table 8.1. Typical values of kinetic parameters for ASM3. ***These values are provided as examples and are not part of ASM3.***

Symbol	Characterization	Temperature 10 °C	Temperature 20 °C	Units
k_H	Hydrolysis rate constant	2	3	$g\ COD_{X_S}\ (g\ COD_{X_H})^{-1}\ d^{-1}$
K_X	Hydrolysis saturation constant	1	1	$g\ COD_{X_S}\ (g\ COD_{X_H})^{-1}$
Heterotrophic organisms X_H, aerobic and denitrifying activity				
k_{STO}	Storage rate constant	2.5	5	$g\ COD_{S_S}\ (g\ COD_{X_H})^{-1}\ d^{-1}$
η_{NOX}	Anoxic reduction factor	0.6	0.6	—
K_{O_2}	Saturation constant for S_{NO_2}	0.2	0.2	$g\ O_2\ m^{-3}$
K_{NOX}	Saturation constant for S_{NOX}	0.5	0.5	$g\ NO_3^-\text{-}N\ m^{-3}$
K_S	Saturation constant for substrate S_S	2	2	$g\ COD_{S_S}\ m^{-3}$
K_{STO}	Saturation constant for X_{STO}	1	1	$g\ COD_{X_{STO}}\ (g\ COD_{X_H})^{-1}$
μ_H	Heterotrophic max. growth rate of X_H	1	2	d^{-1}
K_{NH_4}	Saturation constant for ammonium, S_{NH_4}	0.01	0.01	$g\ N\ m^{-3}$
K_{ALK}	Saturation constant for alkalinity for X_H	0.1	0.1	$mole\ HCO_3^-\ m^{-3}$
b_{H,O_2}	Aerobic endogenous respiration rate of X_H	0.1	0.2	d^{-1}
$b_{H,NOX}$	Anoxic endogenous respiration rate of X_H	0.05	0.1	d^{-1}
b_{STO,O_2}	Aerobic respiration rate for X_{STO}	0.1	0.2	d^{-1}
$b_{STO,NOX}$	Anoxic respiration rate for X_{STO}	0.05	0.1	d^{-1}
Autotrophic organisms X_A, nitrifying activity				
μ_A	Autotrophic max. growth rate of X_A	0.35	1.0	d^{-1}
K_{A,NH_4}	Ammonium substrate saturation for X_A	1	1	$g\ N\ m^{-3}$
K_{A,O_2}	Oxygen saturation for nitrifiers	0.5	0.5	$g\ O_2\ m^{-3}$
$K_{A,ALK}$	Bicarbonate saturation for nitrifiers	0.5	0.5	$mole\ HCO_3^-\ m^{-3}$
b_{A,O_2}	Aerobic endogenous respiration rate of X_A	0.05	0.15	d^{-1}
$b_{A,NOX}$	Anoxic endogenous respiration rate of X_A	0.02	0.05	d^{-1}

Table 8.2. Typical stoichiometric and composition parameters for ASM3. These values are given as examples and ***are not part of ASM3.***

Symbol	Characterization	Value	Units	
f_{S_I}	Production of S_I in hydrolysis	0	$g\ COD_{S_I}\ (g\ COD_{X_S})^{-1}$	
Y_{STO,O_2}	Aerobic yield of stored product per S_S	0.85	$g\ COD_{X_{STO}}\ (g\ COD_{S_S})^{-1}$	
$Y_{STO,NOX}$	Anoxic yield of stored product per S_S	0.80	$g\ COD_{X_{STO}}\ (g\ COD_{S_S})^{-1}$	Equation 5.3
Y_{H,O_2}	Aerobic yield of heterotrophic biomass	0.63	$g\ COD_{X_H}\ (g\ COD_{X_{STO}})^{-1}$	
$Y_{H,NOX}$	Anoxic yield of heterotrophic biomass	0.54	$g\ COD_{X_H}\ (g\ COD_{X_{STO}})^{-1}$	Equation 5.3
Y_A	Yield of autotrophic biomass per NO_3^--N	0.24	$g\ COD_{X_A}\ (g\ N_{S_{NOX}})^{-1}$	
f_{X_I}	Production of X_I in endog. respiration	0.20	$g\ COD_{X_I}\ (g\ COD_{X_{BM}})^{-1}$	
i_{N,S_I}	N content of S_I	0.01	$g\ N\ (g\ COD_{S_I})^{-1}$	
i_{N,S_S}	N content of S_S	0.03	$g\ N\ (g\ COD_{S_S})^{-1}$	
i_{N,X_I}	N content of X_I	0.02	$g\ N\ (g\ COD_{X_I})^{-1}$	The values below are
i_{N,X_S}	N content of X_S	0.04	$g\ N\ (g\ COD_{X_S})^{-1}$	suggested if X_{SS} is used to
$i_{N,BM}$	N content of biomass, X_H, X_A	0.07	$g\ N\ (g\ COD_{X_{BM}})^{-1}$	model VSS rather than SS:
i_{SS,X_I}	SS to COD ratio for X_I	0.75	$g\ SS\ (g\ COD_{X_I})^{-1}$	$0.75\ g\ VSS\ (g\ COD_{X_I})^{-1}$
i_{SS,X_S}	SS to COD ratio for X_S	0.75	$g\ SS\ (g\ COD_{X_S})^{-1}$	$0.75\ g\ VSS\ (g\ COD_{X_S})^{-1}$
$i_{SS,BM}$	SS to COD ratio for biomass, X_H, X_A	0.90	$g\ SS\ (g\ COD_{X_{BM}})^{-1}$	$0.75\ g\ VSS\ (g\ COD_{X_H\ or\ X_A})^{-1}$

Table 8.3. Short definition of model compounds and typical wastewater composition (primary effluent) for ASM3. The value of TKN considers the composition of the different model compounds as indicated in Table 8.2: $TKN = \Sigma C_i \cdot \iota_{2,i}$ *over all compounds* $i - S_{NOX} - S_{N_2}$. $COD_{tot} = 260\ g\ COD\ m^{-3}$, $TKN = 25\ g\ N\ m^{-3}$.

Compounds		Concentration	Units	
Dissolved compounds				
S_{O_2}	Dissolved oxygen	0	$g\ O_2\ m^{-3}$	
S_I	Soluble inert organics	30	$g\ COD\ m^{-3}$	
S_S	Readily biodegradable substrates	60	$g\ COD\ m^{-3}$	
S_{NH_4}	Ammonium	16	$g\ N\ m^{-3}$	
S_{N_2}	Dinitrogen, released by denitrification	0	$g\ N\ m^{-3}$	
S_{NOX}	Nitrite plus nitrate	0	$g\ N\ m^{-3}$	
S_{ALK}	Alkalinity, bicarbonate	5	$mole\ HCO_3^-\ m^{-3}$	
Particulate compounds				
X_I	Inert particulate organics	25	$g\ COD\ m^{-3}$	
X_S	Slowly biodegradable substrates	115	$g\ COD\ m^{-3}$	
X_H	Heterotrophic biomass	30	$g\ COD\ m^{-3}$	The value below is
X_{STO}	Organics stored by heterotrophs	0	$g\ COD\ m^{-3}$	suggested if X_{SS} is used to
X_A	Autotrophic, nitrifying biomass	>0	$g\ COD\ m^{-3}$	model VSS rather than SS:
X_{SS}	Total suspended solids	125	$g\ SS\ m^{-3}$	$100\ g\ VSS\ m^{-3}$

Table 8.4. Stoichiometric matrix of ASM3 based on the stoichiometric parameters in Table 8.2. This matrix is a typical application of ASM3 but ***it is not suggested as a reliable form of ASM3.***

Compound i →	1	2	3	4	5	6	7	8	9	10	11	12	13
j Process	S_{O_2}	S_I	S_S	S_{NH_2}	S_{N_2}	S_{NOX}	S_{ALK}	X_I	X_S	X_H	X_{STO}	X_A	X_{SS}
↓ Expressed as →	O_2	COD	COD	N	N	N	Mole	COD	COD	COD	COD	COD	SS
1 Hydrolysis		0	1	0.01			0.001		−1				−0.75
Heterotrophic organisms, aerobic and denitrifying activity													
2 Aerobic storage of S_S	−0.15		−1	0.03			0.002				0.85		0.51
3 Anoxic storage of S_S			−1	0.03	0.07	−0.07	0.007				0.80		0.48
4 Aerobic growth of X_H	−0.60			−0.07			−0.005			1	−1.60		−0.06
5 Anoxic growth (denitrific.)				−0.07	0.30	−0.30	0.016			1	−1.85		−0.21
6 Aerobic endog. respiration	−0.80			0.066			0.005	0.20		−1			−0.75
7 Anoxic endog. respiration				0.066	0.28	−0.28	0.025	0.20		−1			−0.75
8 Aerobic respiration of X_{STO}	−1										−1		−0.60
9 Anoxic respiration of X_{STO}					0.35	−0.35	0.025				−1		−0.60
Autotrophic organisms, nitrifying activity													
10 Aerobic growth of X_A	−18.04			−4.24		4.17	−0.600					1	0.90
11 Aerobic endog. respiration	−0.80			0.066			0.005	0.20				−1	−0.75
12 Anoxic endog. respiration				0.066	0.28	−0.28	0 .025	0.20				−1	−0.75

9. ASM3C: a carbon-based model

In some countries the chemical determination of COD-Cr in routine analysis is not possible because of the heavy metals (Hg, Cr and Ag) involved in the analytical procedures. The alternative COD-Mn is not very valuable in the context of ASM3 since it greatly underestimates ThOD. This is a severe limitation for the application of models such as ASM3. In order to facilitate the application of ASM3, the Task Group proposes ASM3C as an adapted version of ASM3, where organic state variables are expressed in terms of organic carbon rather than COD. This allows the use of TOC instead of COD-Cr measurements in order to characterize wastewater and activated sludge. Since experience with TOC is rather limited at this time, ASM3C should be used with great care.

9.1 Definition and measurement of fractions of organic carbon

TOC [$M(C)\ L^{-3}$] stands for Total Organic Carbon and is analytically available in wastewater. For samples with suspended solids, careful homogenization of the samples is important. Care has to be taken that sedimentation of coagulating solids does not lead to erroneous results. This is especially relevant if autosamplers are used where the sample is not stirred before it is injected into the TOC analyser.

Depending on the pretreatment of the sample, the TOC analysis can be used to characterize different fractions of wastewater. In relation to TOC the terms DOC (Dissolved Organic Carbon) and POC (Particulate Organic carbon) are sometimes used. By definition TOC = DOC + POC. DOC is measured after filtration of the samples. All soluble organic model compounds (S_I, S_S) may be expressed in terms of DOC.

POC is available from the difference TOC – DOC. POC may be used to characterize some of the particulate organic model compounds in the influent (X_I, X_H, X_{STO}, X_A) but not for others (not X_S, since a fraction of X_S passes through the membrane filters).

For the characterization of activated sludge as a whole, standard estimation of TOC may easily lead to gross analytical errors. Here elementary analysis of a dry washed sample or the use of specialized TOC equipment may be the methods of choice.

Since the fractionation of TOC in DOC and POC is not entirely compatible with the definition of the model compounds (X_S being the problem), this report will use TOC only, in order to express that a compound is measured in terms of organic carbon.

9.2 Transition from ASM3 to ASM3C

Deriving ASM3C from ASM3 is an easy task if the composition of the organic compounds is redefined (COD is replaced by TOC). Introducing new units (g ThOD (g TOC)$^{-1}$ rather than g ThOD (g COD)$^{-1}$) and accordingly new absolute values for the composition parameters $\iota_{ThOD,i}$ for all organic compounds in the composition matrix yields the basis for estimating the unknown stoichiometric coefficients x_j, y_j, z_j and t_j with the aid of Equations 5.1 and 5.2.

Realizing that the values of $i_{ThOD,??}$ = 1 g ThOD (g COD)$^{-1}$ are actually model parameters chosen to be unity in ASM3 (which is based on the assumption that ThOD is identical to the measured COD), the transition of ASM3 to ASM3C is a rather minor one.

Table 9.1 includes the stoichiometric matrix for ASM3C, revised from Table 5.1. Changes include transition from COD to TOC for expressing the organic state variables, introduction of the $\iota_{ThOD,??}$ values in the first row of the composition matrix and adjusting the fixed values for storage compounds (X_{STO}) in the composition matrix.

9.3 Adjusting kinetic and stoichiometric parameters for ASM3C

With the introduction of TOC based compounds in ASM3C the absolute values and the units of some kinetic and stoichiometric parameters must be adjusted. As an example the aerobic yield of autotrophic biomass Y_A in ASM3C may be obtained from:

Y_A(ASM3) has the units

$$[\text{g COD}_{X_A}\ (\text{g NO}_3^-\text{-N})^{-1}]$$

Y_A(ASM3C) has the units

$$[\text{g TOC}_{X_A}\ (\text{g NO}_3^-\text{-N})^{-1}]$$

$i_{ThOD,BM}$(ASM3) has the units

$$[\text{g ThOD}_{BM}\ (\text{g COD}_{BM}{}^{-1})]$$

Table 9.1. Stoichiometric matrix $\nu_{j,i}$ and composition matrix $\iota_{k,i}$ of ASM3C. The values of x_j, y_j, z_j and t_j can be obtained in this sequence from mass and charge conservation (Equation 5.1) and composition (Equation 5.2). The stoichiometric coefficients (f_{S_I}, f_I, Y_{STO} and Y_H) must be based on organic carbon (Table 9.3).

Compound i →	1	2	3	4	5	6	7	8	9	10	11	12	13
j Process	S_{O_2}	S_I	S_S	S_{NH_4}	S_{N_2}	S_{NOX}	S_{ALK}	X_I	X_S	X_H	X_{STO}	X_A	X_{SS}
↓ Expressed as →	O_2	TOC	TOC	N	N	N	Mole	TOC	TOC	TOC	TOC	TOC	SS
1 Hydrolysis		f_{S_I}	x_1	y_1			z_1		-1				$-i_{X_S}$
Heterotrophic organisms, aerobic and denitrifying activity													
2 Aerobic storage of S_S	x_2		-1	y_2			z_2				Y_{STO,O_2}		t_2
3 Anoxic storage of S_S			-1	y_3	$-x_3$	x_3	z_3				$Y_{STO,NOX}$		t_3
4 Aerobic growth of X_H	x_4			y_4			z_4			1	$-1/Y_{H,O_2}$		t_4
5 Anoxic growth (denitrific.)				y_4	$-x_5$	x_5	z_5			1	$-1/Y_{H,NOX}$		t_5
6 Aerobic endog. respiration	x_6			y_6			z_6	f_I		-1			t_6
7 Anoxic endog. respiration				y_7	$-x_7$	x_7	z_7	f_I		-1			t_7
8 Aerobic respiration of X_{STO}	x_8										-1		t_8
9 Anoxic respiration of X_{STO}					$-x_9$	x_9	z_9				-1		t_9
Autotrophic organisms, nitrifying activity													
10 Aerobic growth of X_A	x_{10}			y_{10}		$1/Y_A$	z_{10}					1	t_{10}
11 Aerobic endog. respiration	x_{11}			y_{11}			z_{11}	f_I				-1	t_{11}
12 Anoxic endog. respiration				y_{12}	$-x_{12}$	x_{12}	z_{12}	f_I				-1	t_{12}
Composition matrix $\iota_{k,I}$													
k Conservatives													
1 ThOD (g ThOD)	-1	i_{ThOD,S_I}	i_{ThOD,S_S}		-1.71	-4.57		i_{ThOD,X_I}	i_{ThOD,X_S}	$i_{ThOD,BM}$	3.00	$i_{ThOD,BM}$	
2 Nitrogen (g N)		i_{N,S_I}	i_{N,S_S}	1	1	1		i_{N,X_I}	i_{N,X_S}	$i_{N,BM}$		$i_{N,BM}$	
3 Ionic charge (Mole +)				1/14		$-1/14$	-1						
Observables													
4 SS (g SS)								i_{SS,X_I}	i_{SS,X_S}	$i_{SS,BM}$	1.80	$i_{SS,BM}$	

$i_{ThOD,BM}$(ASM3C) has the units

$$[\text{g ThOD}_{BM}\ (\text{g TOC}_{BM}{}^{-1})]$$

From this we obtain (with absolute values from Tables 8.2 and 9.3):

$$(\text{ASM3C}) = Y_A\,(\text{ASM3}) \cdot \frac{i_{ThOD,BM}\,(\text{ASM3})}{i_{ThOD,BM}\,(\text{ASM3C})}$$

$$= 0.24\ \text{g COD}_{BM}\ (\text{g NO}_3^-\text{-N})^{-1} \cdot \frac{1}{2.8}$$

$$= 0.09\ \text{g TOC}_{BM}\ (\text{g NO}_3^-\text{-N})^{-1}$$

Adjustments are best based on careful analysis of the units of the parameter to be adjusted. Even the magnitude of seemingly dimensionless coefficients such as Y_H must be adjusted.

9.4 Typical kinetic and stoichiometric parameters for ASM3C

Again the following remarks apply to ASM3C as well.

It is the responsibility of the user of ASM3C to determine the concentrations of relevant compounds in the wastewater, as well as the stoichiometric and kinetic parameters, which apply to the specific case to be dealt with. **Absolute values of these parameters are not part of ASM3C.** They are necessary, however, if ASM3C is to be applied to any specific case.

In Tables 9.2–9.5 a set of typical kinetic and stoichiometric parameters and concentrations of model compounds in a primary effluent is provided for convenience. This indicates neither that ASM3C is meant to be reliable with these parameters in any case, nor that these parameters are the state of the art. They are merely presented here as a reference for testing computer code and as a first estimate for the design of possible experiments that may be used to identify these parameters more accurately. In comparison to the values given in Tables 8.1–8.4 these values have been adjusted to TOC units based on typical composition parameters in Table 9.4. There may be some rounding errors resulting in slight deviations in model predictions between ASM3 and ASM3C. The extra decimal is not provided because it is thought to be accurate but rather in order to result in better comparison of predictions relative to ASM3.

Table 9.2 provides a list of typical kinetic parameters, Table 9.3 suggests some typical stoichiometric parameters, Table 9.4 indicates the composition of a typical primary effluent and finally Table 9.5 is a stoichiometric matrix, based on Table 9.1 and the specific values introduced in Table 9.3.

9.5 Modelling pH with ASM3C

Including organic carbon in ASM3C allows modelling the pH value in the different reactor compartments. If bicarbonate is assumed to be

Table 9.2. Typical values of kinetic parameters for ASM3C. These values are provided as examples and ***are not part of ASM3.*** *In comparison to the values given in Table 8.1 these values have been adjusted to TOC units based on typical composition parameters in Tables 8.2 and 9.3. There may be some rounding errors resulting in slight deviations in model predictions between ASM3 and ASM3C. The extra decimal is provided not because it is thought to be accurate but rather in order to result in better comparison of predictions relative to ASM3.*

Symbol	Characterization	Temperature 10 °C	20 °C	Units
k_H	Hydrolysis rate constant	2.3	3.4	g TOC_{X_S} (g TOC_{X_H})$^{-1}$ d^{-1}
K_X	Hydrolysis saturation constant	1	1	g TOC_{X_S} (g TOC_{X_H})$^{-1}$
Heterotrophic organisms X_H, aerobic and denitrifying activity				
k_{STO}	Storage rate constant	2.9	5.7	g TOC_{S_S} (g TOC_{X_H})$^{-1}$ d^{-1}
η_{NOX}	Anoxic reduction factor	0.6	0.6	–
K_{O_2}	Saturation constant for S_{NO_2}	0.2	0.2	g O_2 m^{-3}
K_{NOX}	Saturation constant for S_{NOX}	0.5	0.5	g NO_3^--N m^{-3}
K_S	Saturation constant for substrate S_S	0.6	0.6	g TOC_{S_S} m^{-3}
K_{STO}	Saturation constant for X_{STO}	1.1	1.1	g $TOC_{X_{STO}}$ (g TOC_{X_H})$^{-1}$
μ_H	Heterotrophic max. growth rate	1	2	d^{-1}
K_{NH_4}	Saturation constant for ammonium, S_{NH_4}	0.01	0.01	g N m^{-3}
K_{ALK}	Saturation constant for alkalinity for X_H	0.1	0.1	mole HCO_3^- m^{-3}
b_{H,O_2}	Aerobic endogenous respiration rate of X_H	0.1	0.2	d^{-1}
$b_{H,NOX}$	Anoxic endogenous respiration rate of X_H	0.05	0.1	d^{-1}
b_{STO,O_2}	Aerobic respiration rate for X_{STO}	0.1	0.2	d^{-1}
$b_{STO,NOX}$	Anoxic respiration rate for X_{STO}	0.05	0.1	d^{-1}
Autotrophic organisms X_A, nitrifying activity				
μ_A	Autotrophic max. growth rate of X_A	0.35	1.0	d^{-1}
K_{A,NH_4}	Ammonium substrate saturation for X_A	1	1	g N m^{-3}
K_{A,O_2}	Oxygen saturation for nitrifiers	0.5	0.5	g O_2 m^{-3}
$K_{A,ALK}$	Bicarbonate saturation for nitrifiers	0.5	0.5	mole HCO_3^- m^{-3}
b_{A,O_2}	Aerobic endogenous respiration rate of X_A	0.05	0.15	d^{-1}
$b_{A,NOX}$	Anoxic endogenous respiration rate of X_A	0.02	0.05	d^{-1}

Table 9.3. Typical stoichiometric and composition parameters for ASM3C. These values are given as examples and ***are not part of ASM3.***

Symbol	Characterization	Value	Units	
f_{S_I}	Production of S_I in hydrolysis	0	g TOC_{S_I} (g TOC_{X_S})$^{-1}$	
Y_{STO,O_2}	Aerobic yield of stored product per S_S	0.91	g $TOC_{X_{STO}}$ (g TOC_{S_S})$^{-1}$	
$Y_{STO,NOX}$	Anoxic yield of stored product per S_S	0.85	g $TOC_{X_{STO}}$ (g TOC_{S_S})$^{-1}$	Equation 5.3
Y_{H,O_2}	Aerobic yield of heterotrophic biomass	0.67	g TOC_{X_H} (g $TOC_{X_{STO}}$)$^{-1}$	
$Y_{H,NOX}$	Anoxic yield of heterotrophic biomass	0.58	g TOC_{X_H} (g $TOC_{X_{STO}}$)$^{-1}$	Equation 5.3
Y_A	Yield of autotrophic biomass per NO_3^--N	0.09	g TOC_{X_A} (g $N_{S_{NOX}}$)$^{-1}$	
f_{X_I}	Production of X_I in endog. respiration	0.20	g TOC_{X_I} (g $TOC_{X_{BM}}$)$^{-1}$	
i_{ThOD,S_I}	ThOD content of S_I	2.8	g ThOD (g TOC_{S_I})$^{-1}$	
i_{ThOD,S_S}	ThOD content of S_S	3.2	g ThOD (g TOC_{S_S})$^{-1}$	
i_{ThOD,X_I}	ThOD content of X_I	2.8	g ThOD (g TOC_{X_I})$^{-1}$	
i_{ThOD,X_S}	ThOD content of X_S	3.2	g ThOD (g TOC_{X_S})$^{-1}$	
$i_{ThOD,BM}$	ThOD content of biomass, X_H, X_A	2.8	g ThOD (g $TOC_{X_{BM}}$)$^{-1}$	
i_{N,S_I}	N content of S_I	0.03	g N (g TOC_{S_I})$^{-1}$	
i_{N,S_S}	N content of S_S	0.10	g N (g TOC_{S_S})$^{-1}$	
i_{N,X_I}	N content of X_I	0.06	g N (g TOC_{X_I})$^{-1}$	The values below are
i_{N,X_S}	N content of X_S	0.13	g N (g TOC_{X_S})$^{-1}$	suggested if X_{SS} is used to
$i_{N,BM}$	N content of biomass, X_H, X_A	0.20	g N (g $TOC_{X_{BM}}$)$^{-1}$	model VSS rather than SS:
i_{SS,X_I}	SS to TOC ratio for X_I	2.1	g SS (g TOC_{X_I})$^{-1}$	2.1 g VSS (g X_I)$^{-1}$
i_{SS,X_S}	SS to TOC ratio for X_S	2.4	g SS (g TOC_{X_S})$^{-1}$	2.4 g VSS (g X_S)$^{-1}$
$i_{SS,BM}$	SS to TOC ratio for biomass, X_H, X_A	2.5	g SS (g $TOC_{X_{BM}}$)$^{-1}$	2.4 g VSS (g X_H or X_A)$^{-1}$

Table 9.4. Short definition of model compounds and typical wastewater composition (primary effluent) for ASM3C. The value of TKN considers the composition of the different model compounds as indicated in Table 9.3: TKN = $\sum S_i \cdot \iota_{2,i}$ over all compounds i – S_{NOX} – S_{N_2}. COD_{tot} = 260 g COD m^{-3}, TKN = 25 g N m^{-3}.

Compounds		Concentration	Units	
Dissolved compounds				
S_{O_2}	Dissolved oxygen	0	g O_2 m^{-3}	
S_I	Soluble inert organics	11	g TOC m^{-3}	
S_S	Readily biodegradable substrates	19	g TOC m^{-3}	
S_{NH_4}	Ammonium	15.4	g N m^{-3}	
S_{N_2}	Dinitrogen, released by denitrification	0	g N m^{-3}	
S_{NOX}	Nitrite plus nitrate	0	g N m^{-3}	
S_{ALK}	Alkalinity, bicarbonate	5	mole HCO_3^- m^{-3}	
Particulate compounds				
X_I	Inert particulate organics	9	g TOC m^{-3}	
X_S	Slowly biodegradable substrates	36	g TOC m^{-3}	
X_H	Heterotrophic biomass	11	g TOC m^{-3}	The value below is
X_{STO}	Organics stored by heterotrophs	0	g TOC m^{-3}	suggested if X_{SS} is used to
X_A	Autotrophic, nitrifying biomass	>0	g TOC m^{-3}	model VSS rather than SS:
X_{SS}	Total suspended solids	125	g SS m^{-3}	100 g VSS m^{-3}

Table 9.5. Stoichiometric matrix of ASM3C based on the stoichiometric parameters in Tables 9.1 and 9.3. This matrix is a typical application of ASM3C ***but it is not suggested as a reliable form of ASM3C.***

j ↓	Compound i → / Process / Expressed as →	1	2	3	4	5	6	7	8	9	10	11	12	13
		S_{O_2}	S_I	S_S	S_{NH_4}	S_{N_2}	S_{NOX}	S_{ALK}	X_I	X_S	X_H	X_{STO}	X_A	X_{SS}
		O_2	TOC	TOC	N	N	N	Mole	TOC	TOC	TOC	TOC	TOC	SS
1	Hydrolysis		0	1	0.03			0.002		–1				–2.40
	Heterotrophic organisms, aerobic and denitrifying activity													
2	Aerobic storage of S_S	–0.47		–1	0.10			0.007				0.91		1.64
3	Anoxic storage of S_S			–1	0.10	0.23	–0.23	0.023				0.85		1.53
4	Aerobic growth of X_H	–1.61			–0.20			–0.014			1	–1.47		–0.15
5	Anoxic growth (denitrific.)				–0.20	0.83	–0.83	0.045			1	–1.72		–0.60
6	Aerobic endog. respiration	–2.24			0.19			0.013	0.20		–1			–2.08
7	Anoxic endog. respiration				0.19	0.78	–0.78	0.069	0.20		–1			–2.08
8	Aerobic respiration of X_{STO}	–3										–1		–1.80
9	Anoxic respiration of X_{STO}					1.05	–1.05	0.075				–1		–1.80
	Autotrophic organisms, nitrifying activity													
10	Aerobic growth of X_A	–48.0			–11.3		11.1	–1.60					1	2.50
11	Aerobic endog. respiration	–2.24			0.19			0.013	0.20				–1	–2.08
12	Anoxic endog. respiration				0.19	0.78	–0.78	0.069	0.20				–1	–2.08

the dominating buffer system, pH can be modelled by adding dissolved carbon dioxide (CO_2, S_{CO_2}) and the proton (H, S_{H^+}) as additional state variables and two processes describing the equilibrium (forward and backward reaction) of bicarbonate dissociation, with fast reaction rates (the ratio of these reaction rates is given by the equilibrium constant). Further the additional stoichiometric coefficients for S_{CO_2} can be obtained from conservation of carbon (a fourth conservative). Stripping of S_{CO_2} must be related to aeration, considering the correct Henry coefficients and possible saturation of rising air bubbles with CO_2.

It is recommended to derive stoichiometric coefficients z_j from a charge balance over bicarbonate rather than the proton. This typically allows for faster numeric integration because the turnover of large amount of charge is modelled via the large pool of S_{ALK} rather than the very small pool of H.

Accurate modelling of pH may require expansion of the model to include carbonate, nitrite, ammonium and phosphate buffers as well.

9.6 Limitations of ASM3C

The limitations introduced above for ASM3 apply equally to ASM3C. Since experience with TOC is still rather scarce, extra care should be taken in analytical procedures.

10. Conclusion

ASM3 and ASM3C correct for most of the defects identified in ASM1. The ASM3 models provide a common base for the simulation of nitrogen removing activated sludge systems for chemical oxygen demand as well as organic carbon based characterization of wastewater and biomass. These two characterization possibilities can analytically be approximated by COD and TOC analysis. The structure of the ASM3 models provides sufficient details such that they may be used in an advanced course on biological wastewater treatment as a didactic tool. These models are designed as the core for further development and inclusion of additional processes and states as may become necessary when biological phosphorus removal, chemical phosphorus precipitation, growth of filaments etc. ought to be included.

The systematic notation, based on an array of state variables, a stoichiometric matrix, a composition matrix, an array of process rates and conservation equations made it especially easy to introduce these models and indicate how a COD based model may be transformed into another base (here organic carbon, TOC).

Neither ASM3 nor ASM3C has yet been tested against a large variety of experimental data. It is expected that future improvements of model structure may still be required, especially for the description of the storage phenomena. It is obvious that in the beginning experience with ASM3 might be inferior to experience with ASM1. But as our experience will improve the two models might well prove to be equivalent. ASM3 has the advantage that its structure does not have to be adjusted in order to be applicable even if ammonium or bicarbonate limits microbial activity. Therefore if we report in a publication that a simulation was performed with ASM3 or ASM3C, it may be assumed that the model structures introduced in this report have been applied unchanged.

It is good practice to indicate if model structure has been changed: This would then be a dialect of ASM3 but not ASM3 itself.

11. References

The following limited citations relate to the topic discussed in this report and may be useful to understand the background and the presentation of ASM3. The Task Group would like to apologize for not following the standard rules of citation of scientific work and acknowledges that a vast literature (and communication with peers) has stimulated its work. It appears impossible to explicitly identify the specific source of the elements of ASM3 and ASM3C.

Gujer, W. and Larsen, T.A. (1995) The implementation of biokinetics and conservation principles in ASIM. *Wat. Sci. Technol.* **31** (2), 257–266.

Gujer, W., Henze, M., Mino, T. and van Loosdrecht, M.C.M. (1999) Activated Sludge Model No. 3. *Wat. Sci. Technol.* **39** (1), 183–193. [During publication of this paper some typographical errors were introduced. These were corrected in an erratum: Gujer, W., Henze, M., Mino, T. and van Loosdrecht, M. (1999) Errata: Activated Sludge Model No. 3. *Wat. Sci. Technol.* **39** (12), page I. Additionally, the original of this paper was published in the preprints of the 4th IAWQ Seminar on Modelling and Microbiology of Activated Sludge Processes, Kollekolle, Denmark, 16–18 March 1998.]

Henze, M., Grady, C.P.L. Jr, Gujer, W., Marais, G.v.R. and Matsuo, T. (1987) *Activated Sludge Model No. 1.* (IAWPRC Scientific and Technical Report No. 1.) London: IAWPRC.

Henze, M., Gujer, W., Mino, T., Matsuo, T., Wentzel, M.C. and Marais, G.v.R. (1995) *Activated Sludge Model No. 2.* (IAWQ Scientific and Technical Report No. 3.) London: IAWQ.

Henze, M., Gujer, W., Mino, T., Matsuo, T., Wentzel, M.C., Marais, G.v.R. and van Loosdrecht, M.C.M. (1999) Activated Sludge Model No. 2d, ASM2d. *Wat. Sci. Technol.* **39** (1), 165–182.

IWA Publishing's authorised EU representative for General Product Safety Regulations is Diane D'Arras, 15 rue Duret, 75116 Paris, France, e-mail: safety@iwap.co.uk.

Printed and bound by CPI Group (UK) Ltd, Croydon, CR0 4YY
12/05/2026
02108893-0016